The **Essential** Buyer's Guide

MERCEDES-BENZ W124

All models 1984 to 1997
Saloon, Coupé, Cabriolet and Estate

Your marque expert:
Tobias Zoporowski

For post publication news, updates and amendments relating to this book please visit www.veloce.co.uk/books/V4877

First published by Heel Verlag-GmbH, Gut Pottscheidt, 53639 Königswinter, Germany under the title - "Mercedes-Benz W 124 Alle Modelle 1984-1997."

This edition first published in April 2016 by Veloce Publishing Limited, Veloce House, Parkway Farm Business Park, Middle Farm Way, Poundbury, Dorchester, Dorset, DT1 3AR, England.
Fax 01305 250479/e-mail info@veloce.co.uk/web www.veloce.co.uk or www.velocebooks.com.
English translation and adaptation: Julian Parish. Cover photo: Jost Neßhöver.

ISBN: 978-1-845848-77-4 UPC: 6-36847-04877-8

Typesetting, design and page make-up all by Veloce Publishing Ltd on Apple Mac. Printed in India by Replika Press.

Introduction
– the purpose of this book

"The last real Mercedes ..." With these words, enthusiasts have sent nearly every series of Mercedes produced into (a busy) retirement. The slogan 'The best or nothing' dates back to a time when the Swabian engineers could seemingly build their cars for eternity, with no restrictions. In this sense, 'Car number 124,' is really just that: the last real Mercedes. Some 2.5 million cars were built from 1984 until 1995 (for the coupé and estate until 1996, for the cabriolet or convertible until 1997), and towards the end of its life it became the first 'E-Class.'

For all our enthusiasm for its high-quality design and durability, we shouldn't forget that the youngest survivors of the 'middle-class' series, as Mercedes-Benz called the car which the well-off dreamt of owning, are now 20 years old. Moreover, the W124 is a modern and technically sophisticated car, with various electronic control units for its safety and convenience features, which every now and then need professional help. These should not be neglected if the pleasure of running this classic is to last.

Saloon (sedan), estate (station wagon), coupé and cabriolet – in all its variants, nearly 2.5 million units of the W124 series were produced. (Courtesy Frank Homann)

There is still an enormous choice among four body styles (saloon, estate, coupé and convertible) and an almost endless range of colours and options available today. In the meantime, prices for good cars are increasing quickly. The cabrio is the most expensive 124 model by some margin, the top-selling four-door saloon the best value.

The lifecycle of the 124 range was broken up by a series of face-lifts. The first full update took place in 1989; these models can be recognised, among other changes, by their broad side mouldings. With the second face lift in 1993, Mercedes switched to water-based paints and introduced 16-valve engines. In addition, the range was known as the 'E-Class' from then on. Today, examples of this last generation are the least sought-after W124 models. Problems with the paint on these versions have been the cause of corrosion damage. Well-maintained saloons from this generation are therefore the most affordable way to start this hobby.

Diehard enthusiasts are not so keen on the final generation, keeping prices down, even for good cars. (Courtesy Daimler AG)

As far as the engines are concerned, it's hard to go wrong. The diesels – especially the naturally aspirated units – are capable of astronomical mileages, but aren't very quick. The petrol engines are renowned for their refinement, and with regular maintenance are almost as long-lived as the diesels. Automatic transmissions are generally more popular. It's not a matter of their longevity: both the manual and automatic transmissions are exceptionally strong and normally last the lifetime of the car. The manual gearboxes have long been criticised for being notchy and stubborn. But one thing is for sure: the smooth and, ideally, jolt-free automatics are the best match for the level of comfort expected of a range of cars which are among the highest quality ever produced.

The picture symbolises the current market situation: for now, the saloons are still very affordable; the prices for good estates are going up. (Courtesy Daimler AG)

Contents

The Essential Buyer's Guide™ currency
At the time of publication a BG unit of currency "●" equals approximately £1.00/US$1.42/Euro 1.31. Please adjust to suit current exchange rates using Sterling as the base currency.

1 Is it the right car for you?

– marriage guidance

Suitability for daily use

Any Mercedes W124 is still entirely suitable for daily use today. With one minor reservation: in standard form, the diesels may not meet the regulations for particulate emissions gradually being introduced in many European cities. There are, however, companies which now offer aftermarket catalytic converters and particulate filter kits. With these, even the diesels can pass the emissions tests.

Cabriolet, coupé and a rare limousine line up at a 'Mercedes Festival' outside Paris. (Courtesy Julian Parish)

Interior space

All the W124 body styles offer space for at least four adults and their luggage. Even tall drivers will have no problems. The coupé and convertible are strict four-seaters, and, for the saloon, individual rear bucket seats were available as an option. Otherwise, the saloons (sedans) and estates (station wagons) are suitable for five passengers.

Ease of use

Very little effort is needed to steer the car, thanks to power assistance and the huge, ship-like steering wheel. Once you have become used to the foot-operated parking brake, the controls of a W124 hold no mysteries. All the control lights, instruments and switches are self-explanatory, and even well-equipped models avoid having too many buttons. Only the curious single-arm windscreen wiper, with its technically-interesting sweeping mechanism, is somewhat controversial. Some drivers find it too slow and complain of its tendency to leave smears on the windscreen.

An estate is always ready for a long journey. (Courtesy Daimler AG)

Luggage space

At a pinch, you could spend a weekend away in the luggage compartment of a W124 estate! The estate (T-model) is what it always has been: one of the

A cockpit like this needs no explanation. (Courtesy Daimler AG)

roomiest packhorses on the roads of Europe. The saloon, coupé and convertible also offer very generous and well-lit luggage compartments. There is, however, no through-loading facility to the interior.

The luggage compartment of the T-model was one of the roomiest of its day. To increase the space, the backrests can be folded down separately or together. If you first lift up the seat cushions, there is a flat load surface right through to the front seats. Even a long painter's ladder can then be carried. (Courtesy Daimler AG)

Will it fit in the garage?

Even the really long estate will fit in a standard garage. In terms of its width, a W124 is almost skinny next to a 2014 E-Class.

Running costs

The costs of running and maintaining a W124 are naturally very dependent on the engine you choose. In addition, the early models were not originally equipped with a catalytic converter. All the diesels are pretty easy on fuel, even by today's standards. The big petrol-engined cars – especially the 260E – are fond of a drink. From an economic point of view, a 200E or 230E offer the best combination of a classic Mercedes feel and reasonable costs in keeping with the times.

Parts

Ordered today, delivered tomorrow. That goes for any Mercedes dealer and for pretty much any part you need. Considerably cheaper, and often of comparable quality, there is a good range of parts available from independent stockists. Many suppliers specialise in the model and sell used and new replacement parts at fair prices. The first port of call is Mercedes' own Used Parts Centre (MBGTC), which offers completely overhauled parts with a guarantee. There are also breakers which specialise in Mercedes.

Insurance

If you want to run a 230E as a daily driver, you shouldn't pay more than for a run-of-the-mill Golf.

Meanwhile, many insurers recognise the series as a classic car, for which reduced premiums apply. In this case, however, you may not be able to run the car every day. The powerful coupés and the high-performance 400E and 500E saloons are naturally not as cheap to insure.

Investment potential

Good convertibles, coupés in original condition, and well-equipped estates with a clear service history are increasing markedly in price. Among the still numerous saloons, bargains can be found. Meanwhile, prices for cars in good or very good condition are stabilising. In the short term, a big jump in values is not expected. Nonetheless, a well-maintained W124 should never go down in value!

Plus points

Hardly any car of its time better symbolises the middle-class prosperity of the 1980s than the Mercedes W124. That welcoming feeling is always there. Its design is timeless, its comfort, space and safety still up-to-date. It's not for nothing that you'll find the estates adorning the trendy neighbourhoods of major cities. Maintenance costs are manageable, and spare parts prices too. If a job does need doing, any experienced garage can get a W124 running again.

Foibles

The popularity of the model means that there are plenty of dodgy cars on the market. To get the indestructible naturally aspirated diesels through city pollution tests, you need to allow for a significant investment. Some of the petrol engines (260E, 300E 24V and the 16-valve engines in the final generation) need higher maintenance to deliver good performance. In general, the models built from 1993 are more liable to corrosion than their earlier counterparts.

Alternatives

If you set store by the same virtues which the W124 embodies, but don't need as much space, the W201 (190) might be of interest. Above it – if you want to stick to 'Daimler' – there is only one alternative: the W126 ('S-Class'). If you are looking at the 500E/E500 models, BMW's M5 or the Lotus Carlton/Omega might also be on your list, although these both had manual transmissions. Comparable premium saloons from other manufacturers include the Audi 100, BMW 5-Series, Ford Scorpio, Honda Legend, Opel Senator, Saab 9000 and the Volvo 700 and 900 series.

2 Cost considerations

– affordable, or a money pit?

It should be said from the start that the wide range of models and engines in the 124 series makes it hard to generalise about the running costs incurred. In addition, the calculation will depend – for instance with regard to labour rates – on whether the service or repair work is carried out by an independent garage or an official dealer. All the prices below are therefore only guidelines.

Service (minor/major), without additional work x185/x295
Change accessory drivebelt and tensioners x145
Change timing chain and tensioners x185
Change spark plugs x35-50
Replace thermostat x35
Replace radiator x220
Replace water pump x220-295
New exhaust from catalyser back x255-295
Complete replacement clutch x365
Replace front suspension ball joint x220
Replace front control arm x365
Complete overhaul of rear axle x1100-1470
Replace shock absorbers front & rear (not estate/station wagon) x440
Replace rear shock absorbers (estate) x660
Change both pressure accumulators for self-levelling suspension x330
Exchange wheel bearings (front/rear) x150/260
Change brake pads (front/rear) x90/75
Change brake fluid x45
Universal joint (transmission) x145-185
Replace head gasket from x365
Adjust foot-operated parking brake x35
Wheel alignment (with new steering rods) x145
Replace sunroof frame x735
Replace tailgate closing aid (estate) x700
Overhaul rear wiper mechanism (estate) from x370
Replace engine wiring loom from x735
Replace engine control unit from x735
Replace automatic transmission control unit x880
Replace control unit for ASR (traction control) x880
Air-conditioning: repair leaks, recharge and replace condenser x440
Replace faulty window winder mechanism (front) x220

General guidance

As a rule of thumb, electrical and electronic faults are just as expensive to repair as on a modern car. Occasionally, used or exchange control units can be picked up from online auction sites amazingly cheaply. If you can't spare the money for a new component, it's worth a try, but you may end up with a nasty surprise. As a rule, the seller won't guarantee that the part will work perfectly. Why not? Because the functions of an engine management system can't be tested with everyday tools. As

It's clearly old, but this rear axle and suspension are in good order. A careful look is a must!

The exhaust components are not particularly liable to wear. If, however, a replacement is needed, independent suppliers let you make significant savings.

an alternative, various firms specialise in the repair of control units. Often, the official Mercedes-Benz Classic Center can also help, and even provide a warranty.

Parts that are easy to find

Basically, everything. You're driving a Mercedes-Benz. Which means: whatever your car needs or might need to keep it running, you can obtain either straightaway or within 24 hours from the dealer's parts store. That holds good for parts that wear out too.

Parts that are hard to find

These terms hardly exist in the world of Mercedes. Unless it's a question of interior trim parts in exotic colours – and in 'MB-Tex' or velour at that – you really can get everything. And if your search for a part truly seems hopeless – for example, for a blue dashboard or well-maintained blue seats for the estate, which now and then change hands for astronomical prices – there is always the active enthusiast community. Here someone always has something, or knows someone who does.

Parts that are very expensive

Even if it looks like its lesser-powered brethren, the V8-powered 500E is in a class of its own. That goes for its maintenance requirements and spare parts as well. The motto "If it fits one, it will fit any of them" doesn't apply to this model. In many cases parts will be needed which only fit the 500E. For sure, they are available, but they will be very expensive.

Replacing faulty shock absorbers needn't cost the earth. This one is free from leaks, and still doing its job without any problems.

3 Living with a W124

– will you get along together?

A Mercedes W124 is – at least, in the case of the saloon (sedan) or estate (station wagon) – a classic for the whole family. The author's friends include a couple with three (still quite young) children, who use an estate as their everyday car. Not because they couldn't afford a newer, and therefore much more expensive car, but rather out of conviction. That says a great deal about a range of cars which is so fundamentally down-to-earth and a model of everything Germanic that, even when it is getting on in years, has scarcely any major shortcomings and goes about its daily business reliably and uncomplainingly. Of course, even a W124 can rust, even a W124 occasionally suffers technical failings. Basically, however, faults never arise without being visible. The weak points, whether they affect the body or mechanicals, can almost always be repaired at reasonable cost. Provided that is, that the car has not been completely neglected, in which case you should generally steer well clear. A complete restoration of a W124 isn't worth it. There are still far too many good cars on the market. Which W124 model you decide to live with, will depend on your budget and your personal preferences. As elegant summer cars for stylish cruising, the coupé and convertible are the most suitable. The saloon and estate – whichever you prefer – will accompany you throughout the year. All the models were built over their long production run in so many colours and specifications that even today your choice is almost limitless, and it will come down to personal choice. It's a question of taste. What the Germans call a 'book-keeper' spec 200D in cream, or a fully loaded six-cylinder with automatic transmission, air-conditioning and leather upholstery in metallic grey? Whatever you'd like, you'll find it! It might just take a little longer.

Give yourself enough time and buy the best car you can for your budget. The author is sure of it: with this car you'll grow old together!

The 124 series (the convertible is missing here) is still a familiar sight on our roads. (Courtesy Daimler AG)

An early saloon (from before the first face-lift) in good condition is one for enthusiasts. Soon it will be a fully-fledged classic! (Courtesy Daimler AG)

If there's a car you can load up, this is it! An estate is your leisure companion. (Courtesy Daimler AG)

Getting away from the city for the well-to-do: an A124 (Cabriolet) has always been a pleasure for the happy few. (Courtesy Daimler AG)

4 Relative values
– which model for you?

Saloon

The four-door saloon (sedan) is the 124 series body style which was built in the greatest numbers. Its timeless lines still symbolise the former theme of Mercedes' styling department, which always designed its cars to be just behind the latest fashion. The ideal basis on which to describe it as 'classical.'

Looking for a trouble-free everyday classic? Here it is: a 200E saloon following the first face-lift, with automatic transmission and a sunroof. (Courtesy Frank Homann)

There is still a wide selection of saloons, in every possible condition, colour and specification. Moreover, it is the only model in the range for which all the engines were offered, from the rather sluggish 200 Diesel to the 500E with its beefy V8 engine.

A special mention should be given here to the six-door Limousine models, introduced at the end of 1989. Available with five-cylinder diesel and six-cylinder petrol engines, these were popular with embassies and funeral directors.

Very early cars suffered from quality problems, which Mercedes dealt with rigorously after strong protests from their regular customers (taxi drivers). In the meantime, an early 124 saloon, which gets by without the broad side mouldings characteristic of later cars, is of interest to collectors. From autumn 1985 catalytic converters were introduced, and from autumn 1988 ABS on all models. From 1989 to 1993 the series reached its highest level of quality, which was never surpassed. If you are looking for a W124 not just as a collector, but to use as an everyday car, these are the years on which you should concentrate.

The sweetest temptation since Daimler began production of the 124 series: the timelessly elegant coupé. (Courtesy Daimler AG)

Coupé

Many fans of this series consider the coupé, which was available from March 1987, as its designer Bruno Sacco's masterpiece and one of the most beautiful cars of the late eighties. That is a matter of taste. But it's a fact that the elegant two-door introduced the plastic side cladding to the range; known in German as 'Sacco panels,' the designer himself never appreciated the term. The coupé never existed without them.

As a result, the two-door was only offered as a four-seater. In fact, it is decently comfortable for adults and for long journeys. Its distinctive feature is the side profile without a B-pillar. When all the windows are open – even the small three-quarter windows can be lowered completely – it begins to feel like a convertible.

Like the open version, the coupé is equipped with electric seatbelt guides; if possible, these should still work, as replacements are expensive. Over the years, many coupés have fallen victim to tuning. Shortened and stiffened suspension kits, as well as over-wide tyres, will certainly have spoilt the precious bodywork. Cars like this, including those which have been converted back to their original specification, should be avoided. Even when the price is tempting.

Cabriolet

The open version of the 124 arrived late in the day. The full four-seater convertible was only available from autumn 1991. Like a whirlwind, it conquered the garages of up-market suburbs and the hearts of well-to-do doctors' wives. For the most part, this open-air cruiser – with its enormously rigid construction and multi-layered hood (soft top), which was completely suitable for year-round use – was granted an agreeable existence as a second car. This is reflected today in the number of well cared-for cars on offer, which come at a price.

The open-top 124 never existed with diesel engines, a combination that is commonplace today. Even when ordered with the smallest engine, the cars were more often than not comprehensively equipped. Leather and metallic paint were (almost) compulsory and automatics common. The roof opened and closed electrically, and in most cases should still do so 20 years on. Top safety feature: in the case of a rollover accident, hidden hoops would pop up with lightning speed from behind the rear seat to lessen the consequences of the accident. Whether this feature still works in the car's old age is not something you can, or would want to test. But you can check that the seatbelt guides work, which are expensive to replace.

A collector's item and an investment: the convertibles were always exclusive. (Courtesy Frank Homann)

Estate

In Mercedes' terminology, the estate is not a vulgar utility wagon but rather a 'station wagon' or 'T-model.' Don't forget that when you open the tailgate to the load compartment, which is still huge by today's standards, and ask yourself whether the luggage compartment with its cathedral-like lighting set into the C and D pillars and its deep-pile carpeting isn't too good for paint pots and carpet offcuts. It is!

But, if necessary, even building materials will fit in the long load bay. Or everything which a family of four would need for three weeks' holiday. Odds and ends can be stored in two roomy underfloor lockers and in the so-called 'smugglers' compartments' on each side. The self-levelling suspension – fitted as standard – ensures that the back of the car doesn't give way, while the ingenious combination load cover and divider protects the rear compartment from prying eyes and the car's occupants from luggage flying around.

One option now much sought after by Mercedes enthusiasts with growing families is the folding rear seat in the luggage compartment. This provides rearward-facing seats for two children, complete with seat belts, and can be folded away completely when not in use.

One comment on the self-levelling suspension: if the rear of the car stands unnaturally high when it is unladen, and the

The estates shouldn't have to work hard. They feel most comfortable as stylish transporters for creative types and freelancers. (Courtesy Frank Homann)

rear suspension is conspicuously harsh, it's a sign of faulty pressure accumulators. The enthusiast community has nicknamed them 'bull's eyes' because of their steel grey finish and characteristic shape. If they are replaced, everything usually works fine again. Many garages which are unfamiliar with the model advise owners unnecessarily to change the very expensive shock absorbers, which should be a last resort.

King of the diesels: the OM 603 (300D) with six cylinders. (Courtesy Daimler AG)

Petrol or diesel?

This is above all a question of driving style rather than mechanical durability. With reasonably regular maintenance, the diesels (internal code: OM) seemingly run for ever – as an entire generation of taxi drivers will enthusiastically confirm even today – but are recommended for those of a placid disposition. That applies above all to the basic OM 601 diesel, which produces a heady 71bhp from its 2-litre engine. The smooth 250D (OM 602) is a special tip, with 89, then 94bhp (and later as much as 112bhp in OM 605 Turbo form in the E250D) and its distinctive, mellow sound. The big six-cylinder diesel (OM 603) in the 300D allowed a positively vigorous turn of speed. Later, the OM 606 (E300 Turbodiesel) had 145 healthy horses on hand, and could transport its passengers as quickly as a 230E (2.3-litre M 102 petrol engine). And that was very acceptable. The fuel consumption of these diesels can no longer, of course, be compared with that of today's direct injection engines, but with a suitable driving style, anticipating road conditions, 35-40mpg (Imp)/29-34mpg (US) is possible.

For the coupé, cabrio and estate the M 103 (300E) was the top engine available. (Courtesy Daimler AG)

The four-cylinder petrol engines with two valves per cylinder, which were produced until 1992, are regarded as particularly long-lived and undemanding units, provided they are regularly maintained. The basic 200, with 104bhp (with carburettors) or 116bhp (fuel-injection), will only satisfy the least demanding drivers.

Thanks to its torque output, the next model up, the 230E with 130bhp, is clearly the better alternative, and is considered by many as the best engine in the entire range. This helps explain its wider distribution to this day. All three engines were developed from the M 102 unit.

The type M 119 V8 (400E/500E) won respect for its sports car-like performance. (Courtesy Frank Homann)

With the introduction of four-valve technology, Mercedes-Benz replaced the power units which had been known since the W123 with the new M 111 family of engines (in the 200E/E200 and 220E/E220).

The 260E – based, like the 300E, on the M 103 – is something of an exception, although it is basically a great, vibration-free six-cylinder. The high revs encouraged by its turbine-like operation occasionally lead to head gasket and camshaft problems.

Driven a bit more gently, this power unit is well suited to the 124. This applies as well to the 300, which has a better reputation. The 280E/E280 and 320E/E320, which were offered from the second half of 1993, have the four-valve M 104 engine up front.

The E36 AMG models (coupé, convertible and estate only) are an enthusiast's choice, with a more powerful, 3.6-litre version of the M 104 power unit and uprated suspension.

The type M 119 V8 petrol engines in the 400E/E420 and 500E/E500 are an experience on their own, and give the imposing saloon sports car-like performance. Running costs and fuel consumption, however, are correspondingly high.

The first (pre-face-lift) generation has yet to gain the side mouldings characteristic of later models. (Courtesy Daimler AG)

Trim levels

The W124 was built before Mercedes introduced trim levels like Elegance or Avantgarde, or the myriad packs of special equipment available today. You may, however, find some cars – like the 300E-24 saloon – in Sportline spec. These had no more power, but uprated suspension, and a choice of leather or cloth – in a distinctive check pattern – for the sports seats.

From September 1989 came the colour-matched side cladding with polished trim strips, new seats and more wood inside. (Courtesy Daimler AG)

Face-lift one or two

Here, things are clear. The older cars are the best! Admittedly, the last generation (after the second face-lift) brought clear indicator lenses, tinted rear lights and – thanks to a redesigned grille – a more modern and friendlier looking 'face.' But the switch to water-based paint and 16-valve engines harmed the cars' previous reputation for longevity. Indeed, cars from the last generation rusted more, and more quickly, in their trouble spots. If you can find a well looked after car with a clear service history, built from 1989 to mid-1993 (face-lift one), it should be your preferred choice. The convertible, which was built until 1997, is a special case. The open-air tourers are often in noticeably better condition, regardless of which generation they come from, than their closed counterparts. This is above all due to the fact that they spent much of their life tucked away in garages, and rarely had to endure salted roads, while very few were worn out racking up big mileages.

The last generation (face-lift two) had an updated appearance, and standard equipment was improved, but it suffered from decreasing quality and the water-based paint used, which was not effective in preventing corrosion. (Courtesy Daimler AG)

The following tables give a good overview of the relative values of the different models at the time this guide went to press. The valuations are based on data from Classic Data. The percentages show the approximate prices relative to the most commonly available model, the 230E petrol-engined saloon, as produced from 1984 to 1992, with a value of 100% corresponding to a 230E in good, but not concours condition. In practice, prices can vary depending on the car's specification and history. Originality and condition play the biggest part in determining the prices fetched.

Price tables

W124 petrol-engined saloons

Find the current value of a 100% car in one of the various published and online free guides to make value comparisons.

Model	Production years	Bhp	Cc	Relative value
200	1985-1989	108	1996	70%
200E	1988-1989	120	1996	70%
200E Kat	1989-1992	116	1996	80%
200E Kat	1992-1993	134	1996	90%
220E Kat	1992-1993	148	2199	90%
230E	1985-1989	134	2298	100%
230E	1989-1992	130	2298	100%
260E	1985-1989	164	2597	80%
260E Kat	1986-1989	158	2597	120%
260E Kat	1989-1992	158	2597	120%
280E Kat	1992-1993	194	2799	130%
300E	1985-1989	185	2932	90%
300E Kat	1986-1989	178	2960	110%
300E Kat	1989-1993	178	2960	110%
300E-24 Kat	1989-1992	217	2960	110%
320E Kat	1992-1993	217	3199	110%
400E	1992-1993	275	4196	200%
500E Kat	1990-1993	316	4973	490%
E200	1993-1995	134	1998	90%
E220	1993-1995	148	2199	90%
E280	1994-1995	190	2799	130%
E320	1993-1995	217	3199	110%
E420	1993-1995	275	4196	240%
E500	1993-1995	316	4973	520%
E60 AMG	1993-1994	376	5956	590%

W124 diesel-engined saloons

Model	Production years	Bhp	Cc	Relative value
200D	1985-1989	71	1997	60%
200D	1989-1993	74	1997	60%
250D	1985-1989	89	2497	60%
250D	1989-1993	93	2497	60%
250D Turbo	1988-1989	124	2497	70%
250D Turbo	1989-1993	124	2497	70%
300D	1985-1989	108	2996	80%
300D	1989-1993	111	2996	80%
300D Turbo	1987-1989	141	2996	70%
300D Turbo	1989-1993	145	2996	70%
E200D	1993-1995	74	1997	60%
E250D	1993-1995	111	2497	80%
E250 Turbodiesel	1993-1995	124	2497	70%
E300D	1992-1995	134	2996	80%
E300 Turbo D 4Matic	1993-1995	145	2996	80%

W124 coupés

Model	Production years	Bhp	Cc	Relative value
200CE	1992-1993	134	1998	150%
220CE Kat	1992-1993	148	2199	160%
230CE	1987-1992	130	2298	130%
300CE	1987-1989	185	2960	150%
300CE Kat	1989-1992	177	2932	160%

Model	Production years	Bhp	Cc	Relative value
300CE-24 AMG	1992-1993	249	3314	240%
300CE-24 Kat	1989-1992	217	2960	170%
320CE Kat	1992-1993	217	3199	180%
E200C	1993-1996	134	1998	150%
E220C	1993-1996	148	2199	160%
E320C	1993-1996	217	3199	180%
E36 AMG	1993-1996	268	3595	320%

W124 cabriolets (convertibles)

Model	Production years	Bhp	Cc	Relative value
300CE-24 Kat	1991-1993	217	2960	340%
E200	1994-1997	134	1998	310%
E220	1993-1997	148	2199	330%
E320	1993-1997	217	3199	450%
E36 AMG	1993-1996	268	3595	880%

W124 petrol-engined estates

Model	Production years	Bhp	Cc	Relative value
200T	1985-1990	104	1998	100%
200T	1985-1989	108	1998	60%
200TE Kat	1988-1989	116	1996	100%
200TE Kat	1989-1992	116	1996	110%
200TE Kat	1992-1993	134	1995	110%
220TE	1992-1993	148	2199	120%
230TE Kat	1986-1989	134	2298	70%
230TE Kat	1989-1992	130	2298	100%
280TE Kat	1992-1993	194	2799	100%
300TE24V	1989-1992	217	2960	130%
300TE Kat	1986-1989	178	2960	130%
300TE Kat 4Matic	1989-1993	178	2995	140%
320TE	1992-1993	217	3195	130%
E200T	1992-1996	134	1998	110%
E220T	1993-1996	148	2199	120%
E280T	1993-1996	190	2799	130%
E300T 4Matic	1993-1995	178	2995	140%
E320T	1992-1996	217	3199	140%
E36 AMG	1992-1995	268	3606	280%

W124 diesel-engined estates

Model	Production years	Bhp	Cc	Relative value
200TD	1986-1989	71	1997	70%
200TD	1989-1991	74	1997	70%
250TD	1985-1989	89	2479	70%
250TD	1989-1993	93	2497	70%
300TD	1986-1989	108	2975	80%
300TD	1989-1993	111	2996	80%
300TD Turbo	1987-1989	141	2975	130%
300TD Turbo	1989-1993	145	2996	130%
300TD Turbo 4Matic	1987-1989	141	2975	140%
300TD Turbo 4Matic	1989-1993	145	2995	140%
E250TD	[illegible]	111	2497	120%
E300T Turbo D	1993-1996	145	2996	130%
E300TD	1992-1996	134	2996	130%

5 Before you view

– be well informed

To avoid a wasted journey, and the disappointment of finding that the car does not match your expectations, it will help if you're very clear about what questions you want to ask before you pick up the telephone. Some of these points might appear basic, but when you're excited about the prospect of buying your dream classic, it's amazing how some of the most obvious things slip the mind ... Also check the current values of the model you are interested in in classic car magazines which give both a price guide and auction results.

Where is the car?

Is it going to be worth travelling to the next county/state, or even across a border? A locally advertised car, although it may not sound very interesting, can add to your knowledge for very little effort, so make a visit – it might even be in better condition than expected.

Cars like this are rarely on offer: a one-owner 230E with a complete history, well-equipped and with under 50,000km (31,000 miles). Cars like this don't come cheap, especially from a reputable dealer.

Dealer or private sale

Establish early on if the car is being sold by its owner or by a trader. A private owner should have all the history, so don't be afraid to ask detailed questions. A dealer may have more limited knowledge of a car's history, but should have some documentation. A dealer may offer a warranty/ guarantee (ask for a printed copy), and finance.

A good buy, when you know what you're getting into: a diesel saloon at a fair price from a private seller, here at the Bremen Classic Motorshow.

Cost of collection and delivery

A dealer may well be used to quoting for delivery by car transporter. A private owner may agree to meet you halfway – but only agree to this after you have seen the car at the vendor's address to validate the documents. Conversely, you could meet halfway and agree the sale but insist on meeting at the vendor's address for the handover.

View – when and where

It is always preferable to view at the vendor's home or business premises. In the case of a private sale, the car's documentation should tally with the vendor's name and address. Arrange to view only in daylight and avoid a wet day. Most cars look better in poor light or when wet.

Reason for sale

Do make it one of the first questions. Why is the car being sold and how long has it been with the current owner? How many previous owners?

Left-hand drive to right-hand drive/specials and convertibles

If a steering conversion has been done it can only reduce the value and it may well be that other aspects of the car still reflect the specification for a foreign market.

Condition (body/chassis/interior/mechanicals)

Ask for an honest appraisal of the car's condition. Ask specifically about some of the check items described in Chapter 7.

All original specification

An original equipment car is invariably of higher value than a customised version.

Matching data/legal ownership

Do VIN/chassis, engine numbers and licence plate match the official registration document? Is the owner's name and address recorded in the official registration documents?

For those countries that require an annual test of roadworthiness, does the car have a document showing it complies (an MoT certificate in the UK, which can be verified on 0845 600 5977)?

If a smog/emissions certificate is mandatory, does the car have one?

If required, does the car carry a current road fund license/licence plate tag?

Does the vendor own the car outright? Money might be owed to a finance company or bank: the car could even be stolen. Several organisations will supply the data on ownership, based on the car's licence plate number, for a fee. Such companies can often also tell you whether the car has been 'written-off' by an insurance company. In the UK these organisations can supply vehicle data:

HPI – 01722 422 422
AA – 0870 600 0836
DVLA – 0870 240 0010
RAC – 0870 533 3660

Other countries will have similar organisations.

An E320 estate (station wagon) with original bodykit accessories from AMG and Göckel. (Courtesy Jan Strunk)

An AMG as original as you'll find: here, nothing has been messed about with; all the parts are original and as supplied from the factory. A car like this is a piece of history. (Courtesy Frank Homann)

Unleaded fuel
If necessary, has the car been modified to run on unleaded fuel?

Insurance
Check with your existing insurer before setting out, your current policy might not cover you to drive the car if you do purchase it.

How you can pay
A cheque/check will take several days to clear and the seller may prefer to sell to a cash buyer. However, a banker's draft (a cheque issued by a bank) is as good as cash, but safer, so contact your own bank and become familiar with the formalities that are necessary to obtain one.

Buying at auction?
If the intention is to buy at auction, see Chapter 10 for further advice.

Professional vehicle check (vehicle examination)
There are often marque/model specialists who will undertake professional examination of a vehicle on your behalf. Owners' clubs will be able to put you in touch with such specialists.

Other organisations that will carry out a general professional check in the UK are:

AA – 0800 085 3007 (motoring organisation with vehicle inspectors)
ABS – 0800 358 5855 (specialist vehicle inspection company)
RAC – 0870 533 3660 (motoring organisation with vehicle inspectors)

Other countries will have similar organisations.

6 Inspection equipment

– these items will really help

This book

Before you rush out of the door, gather together a few items that will help as you work your way around the car. This book is designed to be your guide at every step, so take it along and use the check boxes to help you assess each area of the car you're interested in. Don't be afraid to let the seller see you using it.

Reading glasses (if you need them for close work)

Take your reading glasses if you need them to read documents and make close up inspections.

Torch

A torch with fresh batteries will be useful for peering into the wheelarches and under the car.

Magnet (not powerful, a fridge magnet is ideal)

A magnet will help you check if the car is full of filler, or has fibreglass panels. Use the magnet to sample bodywork areas all around the car, but be careful not to damage the paintwork. Expect to find a little filler here and there, but not whole panels. There's nothing wrong with fibreglass panels, but a purist might want the car to be as original as possible.

Probe (a small screwdriver works very well)

A small screwdriver can be used – with care – as a probe, particularly in the wheelarches and on the underside. With this you should be able to check an area of severe corrosion, but be careful – if it's really bad the screwdriver might go right through the metal!

Overalls

Be prepared to get dirty. Take along a pair of overalls, if you have them.

Mirror on a stick

Fixing a mirror at an angle on the end of a stick may seem odd, but you'll probably need it to check the condition of the underside of the car. It will also help you to peer into some of the important crevices. You can also use it, together with the torch, along the underside of the sills and on the floor.

Digital camera (or smartphone)

If you have the use of a digital camera, take it along so that later you can study some areas of the car more closely. Take a picture of any part of the car that causes you concern, and seek a friend's opinion.

A friend, preferably a knowledgeable enthusiast

Ideally, have a friend or knowledgeable enthusiast accompany you; a second opinion is always valuable.

7 Fifteen minute evaluation
– walk away or stay?

In a quarter of an hour you won't be able to judge whether the pile of metal in front of you will be the partner for life you've long been looking for. Some faults won't be seen by the smartest or most serious buyers, others develop insidiously. After all, we're talking here about a car which is at least 20 years old. In any case, you're right to assume that the car will never be in perfect condition. If the car is sold as such, then you can be sure it isn't. The sections below will at least give you a few clues as to whether a closer examination of the object of your desires is worthwhile.

This car looks spotless, but under its polished bodywork it needs work.

Bodywork

Minor parking scrapes like this can often be put right with 'SMART' repairs. (Courtesy Frank Homann)

It should go without saying that you should look at a car in dry and preferably sunny weather. Only then will you be able to see at once if part of the body has been repainted. Differences in colour between the doors, wings and bonnet (hood) – however tiny – wouldn't have been there when the car was first delivered! Nor would wide or irregular panel gaps. There doesn't have to be a serious accident behind it, but the seller should be able to give you a plausible explanation. Grazes on plastic bumpers, spoilers or side mouldings may well just be the battle scars of a car which has been used every day. If, however, these components have been ripped, broken or noticeably deformed, you should get to the bottom of the matter. It's possible that the real – and much greater – damage is hidden underneath it.

This jacking point on a pre-face-lift car is already showing clearly the onset of corrosion. An advantage of the early models without side mouldings is that you can see rust easily and treat it in good time. (Courtesy Frank Homann)

Next, it makes sense to turn to the potential rust spots: look first at all four jacking points. For cars with side mouldings (from the first face-lift onwards), you must remove the covers. If the jacking points show clear signs of rust, you can assume that the sills will also soon be affected. Unless you know an expert welder, you can stop your inspection here: the repairs will be expensive. If the front wings (especially where they meet the bumpers, particularly on the right-hand side), the rear wheelarches (and what the Germans term the 'knee piece' moulding just ahead of them) are already blistering, it will quickly be obvious. The door bottoms and frames (remove the seals!), as well as the tailgate, the frames of

On cars from the first face-lift onwards, you must remove the covers from the jacking points to inspect them. This point is (still) in decent condition.

A wing panel like this can only be replaced. It has rusted through completely from inside. It is essential to check how it looks from behind.

This windscreen frame needs work. And quickly! (Courtesy Frank Homann)

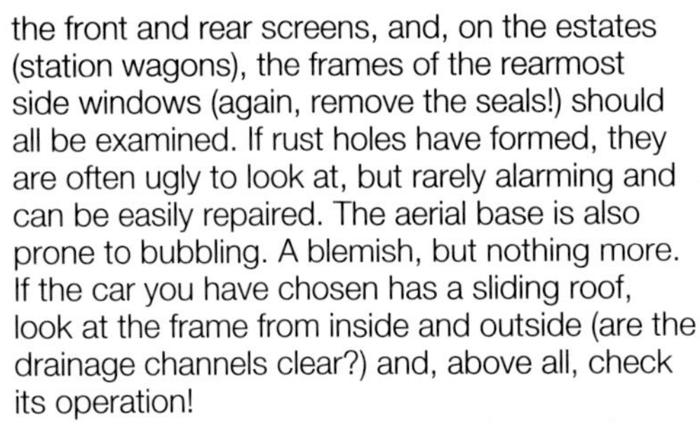

the front and rear screens, and, on the estates (station wagons), the frames of the rearmost side windows (again, remove the seals!) should all be examined. If rust holes have formed, they are often ugly to look at, but rarely alarming and can be easily repaired. The aerial base is also prone to bubbling. A blemish, but nothing more. If the car you have chosen has a sliding roof, look at the frame from inside and outside (are the drainage channels clear?) and, above all, check its operation!

If you have the opportunity to inspect the underside of the car on a lift or over an inspection pit, then do so. Take a look, preferably with a powerful lamp, at the supports for the front springs. Are the springs themselves intact? Sometimes the first coil is broken. Then check the attachment points for the rear shock absorbers and the overall condition of the front and rear axles. Naturally all the fuel, hydraulic and brake pipes running under the car should be free from faults, while the flexible brake hoses should not be cracked or discoloured (light grey). If the underside is conspicuously clean or has had fresh underseal applied, you are right to be suspicious. At the same time, you'll usually notice whether the exhaust system looks sound. Finally, it does

A serious fault seen during technical inspections: broken front springs (the first coil) are unfortunately typical and go unnoticed by the layman, as the handling of the car is scarcely affected.

If it's bubbling outside, then it's rotten inside. Corrosion resulting from contact between the aluminium trim strip and the windscreen frame itself. (Courtesy Frank Homann)

With regular maintenance of the runners, the sliding roof will open and close smoothly and quietly.

No fault found: no work is needed on these lines.

Also typical: worn-through bolsters on the outside edge of the driver's seat appear relatively early on cars with cloth interiors. Velour, leather and 'MB-Tex' are hardly affected by this.

no harm to lift the bonnet and to check the area around and underneath the screenwash reservoir, and – if possible – look at the battery tray.

The wind deflector should pop up and fold down quietly and quickly. A grease spray works wonders!

Interior

When you open the driver's door, the first thing you'll see will be the outside bolster of the driver's seat. That's a good thing, because often it is worn smooth or right through. That doesn't necessarily mean that the car is worn out. Pay attention as well to signs of wear on the pedals and the gear knob. For man-made materials, ask yourself: is the characteristic pattern of the cloth still visible, or has it been rubbed smooth? For leather upholstery, take into account the stitching. Does the material appear greasy or cracked? Does your overall impression of the interior correspond to the car's mileage and year? Are the seat cushions still fairly firm or do they feel worn out? Are there any water stains (be sure to look under the front carpets as well!).

It's impossible to tell that this gear knob has already seen 25 years, and more than 280,000km (170,000 miles) service.

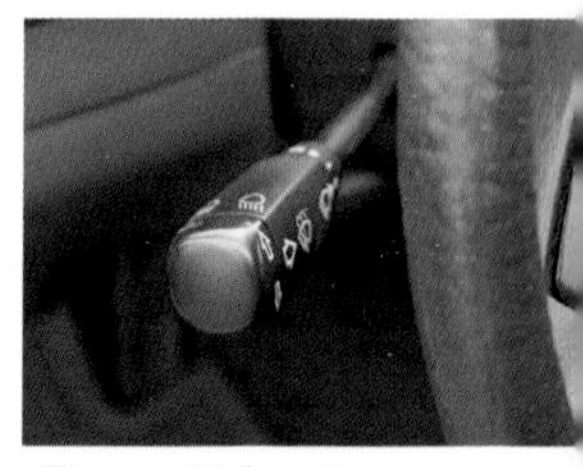
The multi-function column stalk is a sought-after replacement part, which is hard to find used, and anything but cheap to buy new.

For cloth interiors, the sun can often cause fading to the rear parcel shelf or top of the rear seats, even with quite young cars. You can either put up with it or repair it later. It isn't a 'show-stopper' for the car in question. Take in the interior and check the technical features. Do all the buttons and switches work? What about the heating, ventilation, air-conditioning and window winders (even the manual winders can stop working)? It's also important that the sliding roof operates correctly: does it open and close quietly and reasonably fast? Does the tilt function work? Does the wind deflector pop up and down without any problems? Look at the runners for the roof; a thorough clean and spray with grease has already saved many owners a costly repair job.

The light switch is usually hard-wearing.

At some stage the lacquer will peel. Cars which have led a demanding life may suffer worn-out switches, but these can easily be changed.

It's almost usual to find that some of the dashboard lighting on a W124 – whether for the main instrument panel or for the heating and ventilation controls –

has stopped working. That shouldn't give cause for concern, it can be repaired by proficient home mechanics. The bulbs have plug-in sockets and should therefore not be soldered. Other potential sources of trouble lurk in the multi-function column stalk for the indicators, main beam headlamps and windscreen wipers. Replacements don't come cheap. This also applies to the single-arm wiper mechanism. If it operates noticeably slowly or fails to lift during the wiping process, it's a sign of a repair coming soon. Faulty electric radio aerials can easily be replaced by mechanical ones.

The wiper motor itself is basically trouble-free, as long as the lifting mechanism is in good order. If not, a replacement can be pricey.

Mechanical components

With the car on a lift, you should look first at the age and condition of the tyres, the universal joints and propshaft, as well as the brake discs and pads. Open the bonnet and check the coolant and the oil filler cap. Are there any traces of oil in the coolant tank, does it smell of oil or as if something has burnt? Are there signs of foam or white marks on the underside of the oil filler cap? If the answer to one of these questions is 'Yes,' you can be sure that the head gasket has failed or will do so soon.

Look for leaks in the cooling system. Light-coloured marks on the hoses, hose connectors or radiator are a sign of leaks in the cooling system. If there is a strong smell of petrol in the engine compartment, the fuel lines may be weeping. Engines which are completely covered in oil are just as suspicious as spotless engines which have covered 300,000km (nearly 200,000 miles). If the engine looks like it's been working, but shows no signs of leaks, as a rule it will be in good health.

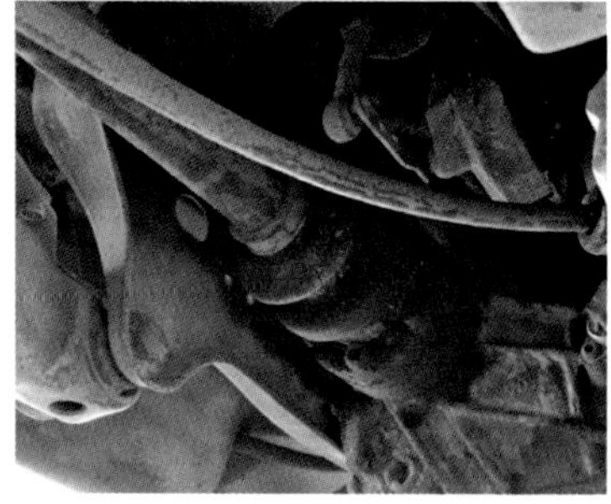

If the rear differential is as free from leaks as this, with no noticeable noises, then usually it will be fine.

Now it's a good idea to start the engine and listen to it running, both from a cold start and after it has warmed up during a test drive. Does it run sweetly and without any noticeable rattling or knocking sounds? Is anything rubbing? Does the engine 'stutter' when idling? Noises from the accessory drivebelt are nothing suspicious. Even after the belt and all the tensioners have been changed, the mechanism can start squeaking again after quite a short time. You just have to live with it. If the tension is OK and the belt looks to be in good condition, as a rule you shouldn't have to worry.

The neck of this radiator has become completely oxidised over the years. What matters more is the colour of the coolant and how it smells.

Someone spilled some oil here when filling the engine, but otherwise it's free from leaks. Conspicuously clean or very dirty (oily) engines are a warning sign.

This accessory drivebelt was replaced, together with its tensioners, not long ago, but still doesn't run completely silently.

With regular use, the foot-operated parking brake causes little trouble.

During the test drive, listen carefully. Do the wheels rumble, or is there an audible noise when cornering? Either a wheel bearing is on its way out, the alignment is out, or the wheels are out of balance. Nothing which can't be fixed. Or rather, you shouldn't wait too long. If your prospective purchase is an automatic, the gears should change smoothly (the automatic gearbox is not completely jolt-free). If you have the feeling that the transmission is taking an abnormally long time to engage the next gear, then usually something is wrong. Sometimes, it's enough to change the transmission oil and clean the filter element. According to many people, both jobs should be done regularly. The car is sure to drive better afterwards. The manual gearboxes have a reputation for being generally rather notchy, but tough. Typically for a Mercedes, you should check the foot-operated parking brake. The pedal travel should not be too long, and it should hold the car securely on a slope. Last of all, check for any free play in the steering. To some extent, it's normal for this to increase in a Mercedes which is getting on in years. But the steering should not seem excessively vague.

Summary

You've probably needed more than 15 minutes to check all these points. But it's time well spent, which will save you cash if you have any doubts.

8 Key points
– where to look for problems

Many of the trouble spots which are critical in determining the condition of a W124 have already been mentioned. Here is a concise summary of the most important points.

Bodywork

Corrosion in and around the jacking points (remove the covers on post-face-lift cars) weakens the body structure, as in most cases the sills will already be affected. Rust can also be found in the frames of the front and rear screens. If the screens have turned milky white at their edges, this is delamination caused by frame corrosion. Check the door bottoms and under the seals on the door frames; on estates (station wagons), examine the lower frame of the side windows (lift the sealing strips from inside). Rust can wreak havoc there undetected. If the frames of the rearmost side windows are already showing blisters from the outside, then substantial welding work will be unavoidable. Rust can also flourish under the side mouldings of cars from the first face-lift onwards. If blisters can be seen on the upper edge of the cladding, the metal underneath has usually had it. Other points to check: the front and rear wheelarches, as well as the so-called 'knee piece' moulding ahead of the rear wheelarch and the area where the front right-hand wing and bumper meet and, last of all, the area around the radio aerial base.

Milky edges like this don't go down well with vehicle test inspectors. The cause is usually to be found in the screen surround.

Here, the corrosion is only superficial and can be removed. But where did it come from? Is water getting into the door or behind the side mouldings because of a failed seal?

Front of the car

How serious are stone chips on the bonnet (hood), bumper moulding and radiator grille? Is the chrome surround of the radiator grille in good condition? Are any of the plastic slats broken? Does the tongue to open the bonnet pop out okay? Is the three-pointed star crooked or, worse still, broken off? Are the headlamp lenses and reflectors clear? Does the single-arm windscreen wiper return to its resting position? Does it lift up cleanly to reach both top corners of the windscreen?

Check whether the heavy bonnet can be safely held in both possible opening positions. Watch out: don't catch your fingers!

Rear of the car

Examine the underside of the boot (trunk) lid or tailgate and the areas around the handle, lock and number plate lights for signs of rust. Do the number plate lights work? Is the warning triangle still clipped into its support and in good order? On the estate, do the

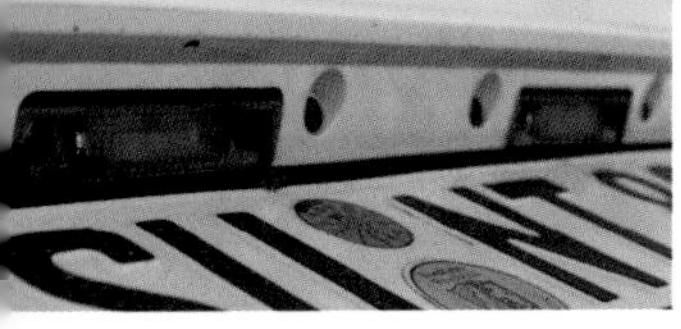

If a bulb here is faulty, changing it takes just a minute. The covers are held in place with screws.

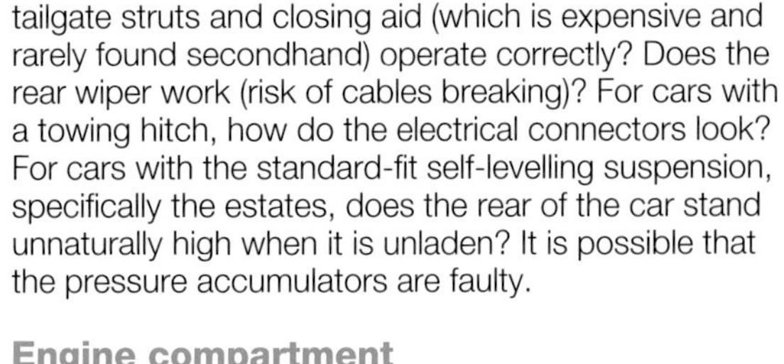

tailgate struts and closing aid (which is expensive and rarely found secondhand) operate correctly? Does the rear wiper work (risk of cables breaking)? For cars with a towing hitch, how do the electrical connectors look? For cars with the standard-fit self-levelling suspension, specifically the estates, does the rear of the car stand unnaturally high when it is unladen? It is possible that the pressure accumulators are faulty.

Not a pretty sight, but no big deal. This sound deadening has been completely eaten away and should be replaced. Some test inspectors will mark it down as a minor fault. (Courtesy Frank Homann)

Engine compartment

Is the engine noticeably dirty or too clean for the mileage it has covered? Neither is good. A light film of oil around the filler is normal; bigger splashes aren't. Is the coolant oily or does it have a burnt smell? Is the inside of the oil filler cap frothy or sticky and whitish? If it is, it may be indicative of head gasket failure. Do light-coloured stains on the cooling system hoses and radiator point to coolant loss? Is the sound deadening under the bonnet in good condition, or has it been eaten away over the years? Are all the fluid reservoirs (especially for brake fluid and the hydraulic fluid for the steering and rear axle) well filled with fresh fluids and leak-free? Look out for loose cables and hoses.

Interior

Does the condition of the interior match the age and claimed mileage of the car? Is the side bolster on the driver's seat next to the door threadbare? Is the steering wheel rim shiny from wear? Are the pedal rubbers and gear knob worn? Have the symbols on the buttons and switches been worn smooth? Does the dashboard lighting work correctly? Do the warning lights and special features, like the optional trip computer or balance control for the rear speakers, all work as they should? Are there any water stains on the headlining (for cars with a sliding roof) or in the foot wells?

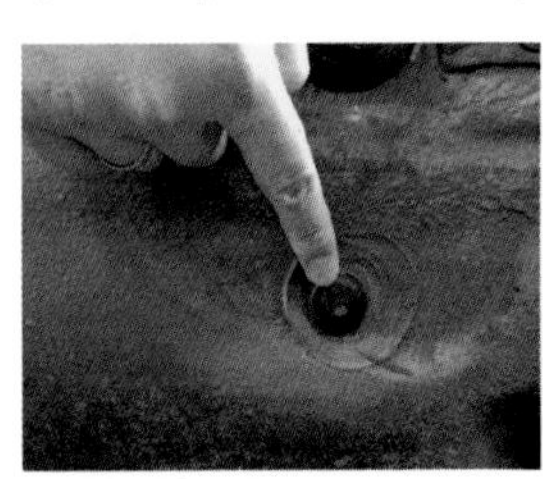

Rust can take hold around rubber plugs like this, which the car has a few of. These should not be under-estimated, as the 'brown plague' usually remains hidden for a long time under the thick and sticky underseal.

Underside

Check the condition of all the fuel, brake and hydraulic lines. Are the ball joints, control arms, front springs and supports free from faults? What about the rear wheel bearings, thrust bearings, tie bars and shock absorbers? Is the degree of wear for the brake pads and discs (rotors), and the age and wear of the tyres, still acceptable? Inspect the universal joints and propshaft. What's the condition of the exhaust system and catalytic converter? Examine the overall state of the underside. Conspicuously thick or freshly applied underseal should arouse your suspicions.

9 Serious evaluation

– 60 minutes for years of enjoyment

Score each section using the boxes as follows: 4 = excellent; 3 = good; 2 = average; 1 = poor. The totting-up procedure is detailed at the end of the chapter.

Be realistic in your marking, and don't let yourself be blinded by your initial excitement when you stand in front of a car which is supposedly in top condition. Keep a cool head and you'll rarely buy a pile of rubbish. Take your time and examine the car thoroughly. If you are careless and hurried, you'll often be punished with expensive repairs. Our tip: bring a magnet, to discover areas of filler. A torch, a tyre tread depth gauge, a screwdriver and a small mirror are also helpful.

Exterior

First impressions

4 3 2 1

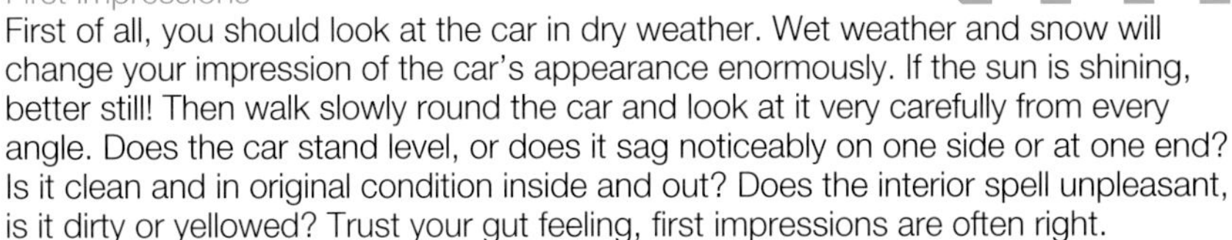

First of all, you should look at the car in dry weather. Wet weather and snow will change your impression of the car's appearance enormously. If the sun is shining, better still! Then walk slowly round the car and look at it very carefully from every angle. Does the car stand level, or does it sag noticeably on one side or at one end? Is it clean and in original condition inside and out? Does the interior spell unpleasant, is it dirty or yellowed? Trust your gut feeling, first impressions are often right.

Colours

4 3 2 1

From today's perspective, there are some pretty funny combinations of colours in the 124 series. Many of them have survived, and even the oddly magnificent combination of rosewood metallic paint and a medium-red velour interior won't necessarily guarantee a price reduction. Quite the opposite: the so-called 'wrong' colours, including the blue-green 'Beryl' metallic and the quite common 'Almandine' red metallic, now have their own fans. In any case, if you don't like a colour, just keep looking. There is still plenty of choice on the market.

A well looked-after W124 is a wonderful everyday companion. (Courtesy Daimler AG)

'Almandine' red is seen relatively often, and is considered rather garish among enthusiasts. (Courtesy Daimler AG)

Eye candy: you have to be self-assured to go for a car in 'Beryl' metallic. (Courtesy Daimler AG)

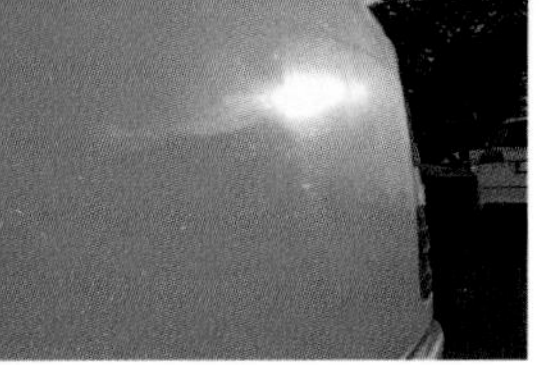

Even solid colours can go dull over time. A good polish can restore some of the shine.

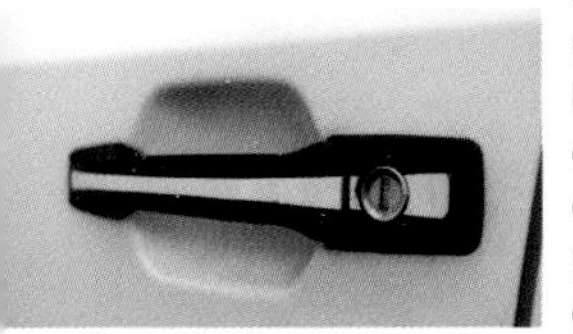

The paint is often scratched around the door handles.

In case of doubt, you can change shabby doors. With a bit of luck, you'll find a replacement in the right colour.

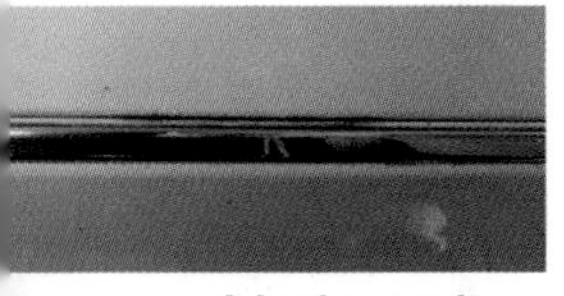

A bad case: it looks harmless enough at the edge of the protective moulding, but when the moulding is removed, significant corrosion of the door may be revealed.

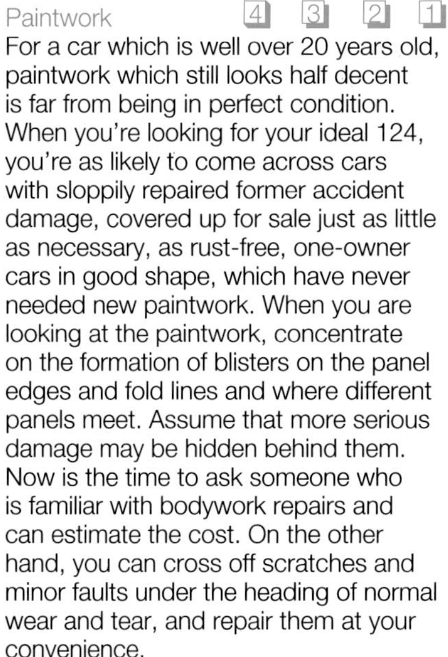

Paintwork 4 3 2 1

For a car which is well over 20 years old, paintwork which still looks half decent is far from being in perfect condition. When you're looking for your ideal 124, you're as likely to come across cars with sloppily repaired former accident damage, covered up for sale just as little as necessary, as rust-free, one-owner cars in good shape, which have never needed new paintwork. When you are looking at the paintwork, concentrate on the formation of blisters on the panel edges and fold lines and where different panels meet. Assume that more serious damage may be hidden behind them. Now is the time to ask someone who is familiar with bodywork repairs and can estimate the cost. On the other hand, you can cross off scratches and minor faults under the heading of normal wear and tear, and repair them at your convenience.

Bodywork 4 3 2 1

No 124 series car left the factory with uneven panel gaps or drooping doors or bonnets (hoods)! They were painstakingly assembled. Bonnets and doors which close poorly, or crooked panel gaps should always make you investigate further. Nearly always, these will have been caused by previous accident damage, which should be used in every case as the basis for a significant reduction in price. For cars which have covered a high mileage, stone chips will be present around the headlamps and on the bonnet. If not, it's possible that the car has been resprayed. In this case too, ask the reason for it. Scratches around the door handles and locks, on the other hand, are quite normal.

Underside 4 3 2 1

If there is a lift or inspection pit available when you are looking at a W124, you should definitely take advantage of it! Have a screwdriver, torch and mirror to

If it's just creaking, lubricating it can help. At some point, however, the retaining straps will have to be replaced.

Water tends to collect between the boot sill and bumper and dries poorly. Rust can start by creeping in here, but then turn nasty. This car is in good shape.

The sill plates often come unstuck as they get old. The start of corrosion underneath isn't always the cause of this, but can be the reason.

hand. Even if everything looks great at first sight, the very thick and tremendously tough PVC underseal applied at the factory can hide a good deal of damage for a long time.

If you notice any inconsistencies on the underside – they might be small blisters or traces of surface corrosion – press these spots carefully or tap them with a screwdriver. In all probability, rust will already have begun its destructive work and you'll feel it crunch.

That isn't always a death sentence for the car in question, but some repairs can be very costly. These include, among others, the rear axle mountings (beware: if there is damage here, the car is no longer safe to drive! Don't be talked into thinking this isn't a serious defect) and the brackets of the front suspension control arms, which can rust through on the cross-members. In this case, you should clarify whether a repair is even possible. Many vehicle testing stations are rightly none too keen on welded repairs to subframes.

The drainage channels in the sills on both sides of the car can be clearly seen (for cars from the first face-lift onwards, take off the side moulding if you have any doubts). Right next to these are the jacking points, which can be examined on cars built from September 1989 (first face-lift with side moulding panels) by opening the plastic caps in the moulded panel. Corrosion damage is common here. If the rust has not gone too far, scruffy jacking points can be refurbished relatively well; repair plates are also available.

Be sure to look under all the wheelarches and – if they are still present – at the rubber plugs. If it looks like rot is setting in here, water has probably been trickling into the sills for some time, where it will cause considerable oxidation. If you don't trust your own judgement of the underside, make sure to bring an expert with you to view the car!

This drainage channel is sound. If blisters are forming around it or if the underseal is loose, it may be hiding nasty corrosion damage.

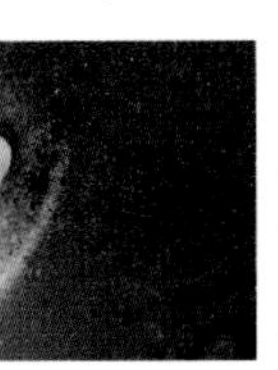

The condition of the wheelarch liners will also provide information about the well-being of the bodywork. Can you see any ominous brown discolouration around these plugs?

Don't just open the cap over the jacking points in the side mouldings, but also look at the support for the jacking plate under the car. If you hear this crunch, alarm bells should sound!

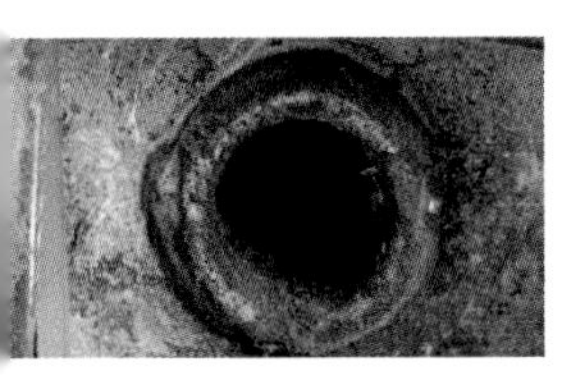

A look behind the jacking point caps (four off) is essential when viewing any 124 from face-lift one onwards.

When the aluminium trim strip shows blistering like this, the window frame underneath it is in bad shape. (Courtesy Frank Homann)

Weather seals 4 3 2 1

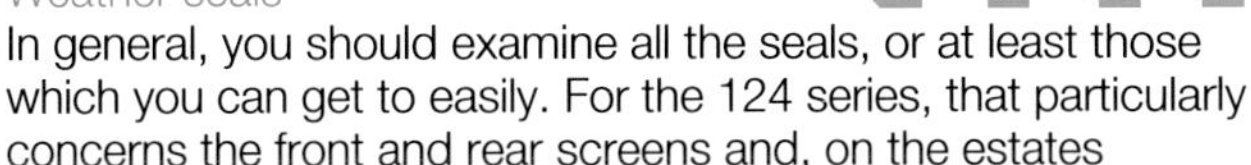

In general, you should examine all the seals, or at least those which you can get to easily. For the 124 series, that particularly concerns the front and rear screens and, on the estates

Milky edges to the windows mean time-consuming and expensive repairs.

Rust like this can be repaired, but it's certainly high time! In this case, the glass must be removed and the frame welded, treated and repainted.

This radiator is a bit dirty, but undamaged and free from leaks. Look out for damage to the core, or white stains.

(station wagons), the rearmost side windows. If you see milky edges to the glass or, on the estates, rust bubbles on the trim strips around the windows, the seal has no longer been doing its job for some time, and a repair will soon be advisable. Repairing the rear side window frames on the estates and the rear screen of the coupé will be especially expensive. All the seals around the bonnet in the engine compartment also demand close scrutiny.

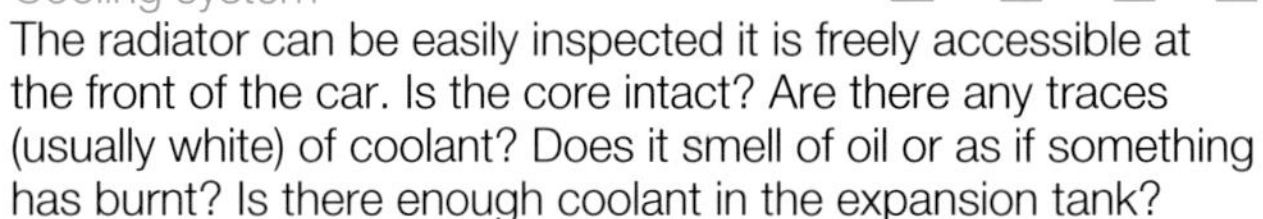

Cooling system

The radiator can be easily inspected it is freely accessible at the front of the car. Is the core intact? Are there any traces (usually white) of coolant? Does it smell of oil or as if something has burnt? Is there enough coolant in the expansion tank?

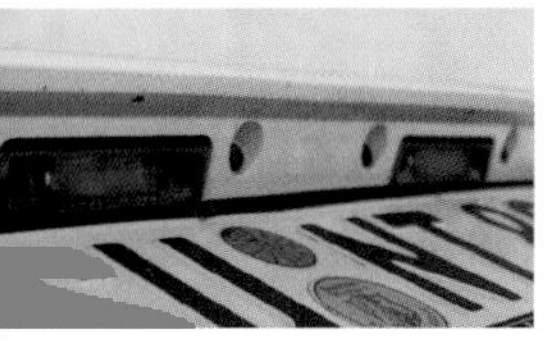

Faulty number plate lights will be noted as a minor defect during official inspections. It's better to change them.

Lights

4 3 2 1

Check all the lights work, including the headlamps, rear lights, indicator, brake lights and fog lights. The rear number plate illumination should also work: if it doesn't, it will be considered a defect during an MoT (in the UK) or similar official test. Models from the final generation also had a third, high-level brakelight.

Windscreen wipers

Apart from the condition of the wiper blades (are they cracked or hard?), which in case of doubt can quickly be replaced, on the 124 series it is the correct operation of the wiper mechanism which should above all be checked. The single-arm wiper moves up in two places, in order to clean the driver's field of view in the uppermost corners of the windscreen. Sometimes, the wiper only works extremely slowly at all speed settings or the lifting mechanism only functions intermittently. Lubricants such

If the rear wiper packs up, it's usually not due to the wiper itself. Often, a broken wire causes it to stop working. The repair is straightforward, but long-winded and therefore not cheap.

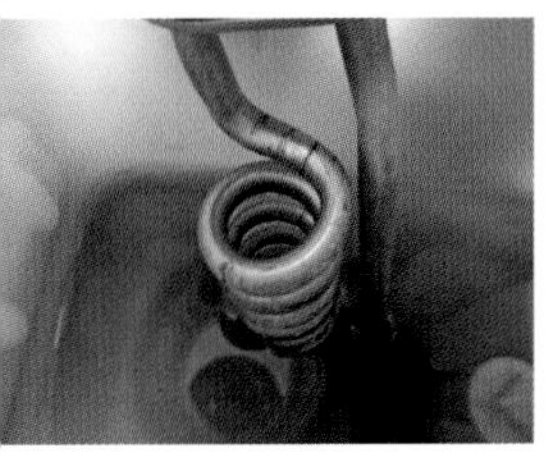

The renowned heating element prevents the washer fluid from freezing. It rarely breaks down.

as WD40 have often worked wonders on such occasions. If the mechanism is faulty, however, it will be expensive. The rear wiper on the estate is often inoperative, which in many cases is due to a broken wire. Check for this by slowly opening and closing the tailgate with the wiper running. If the wiper judders when you are doing this, the diagnosis is clear. To repair the wiring loom, the headlining must be dismantled, which is uncomplicated but time-consuming. The labour charged by a skilled workshop can be correspondingly high.

The rear washer jet position can go wrong, and spray everywhere, not just over the screen.

Windscreen washers 4 3 2 1

Check the condition and operation of the windscreen washers and headlamp washers (if fitted). There are scarcely any problems known with these, but the heating element for the washer fluid in the reservoir should be in working order.

A faulty sliding roof can prove expensive. So be sure to check it all works!

Sunroof 4 3 2 1

Many 124s were delivered from the factory with a sliding roof. Check that this works by opening and closing it several times. Does it operate along its runners quietly, without juddering and reasonably quickly? Does the wind deflector pop up? Does it fold down? Look at the condition of the runners from outside the car with the roof open. Is the sunroof frame clean, and free from leaves and oily smears? Are the drainage channels clear? These can be checked with a length of wire or an old speedometer cable. Does the headlining around the sunroof show signs of mould, or is it clean?

If everything here is clean, more often than not it will operate smoothly.

Soft top 4 3 2 1

The soft top on the convertible is designed for year-round use. Nevertheless, the passage of time can leave its mark, in the form of tears in the folds or makeshift repairs after an attempt to break into the car, because the professional repair of the hood (soft top) was too expensive for the previous owner.

The soft tops are considered unproblematic. However, it shouldn't be ruled out that something might go on strike as the car grows old. Repairs to the complicated folding roof are rarely cheap. (Courtesy Frank Homann)

Which gives you room to negotiate! The folding mechanism and fasteners should operate without excessive force or too much noise. Numerous sensors, limiting switches and valves contribute to the operation of the hood. They should really last for a long time, but once again, keep your eyes and ears open.

Window winders

4 3 2 1

Check the operation of all the window winders (mechanical and electric). With the good old manual winders, the cable mechanism occasionally breaks (particularly on the driver's door); with electric windows, at best it may just be the fuse which is faulty – or the fuse may have worked loose in its holder, something which happens from time to time in the first face-lift series of cars, which were still equipped with the antiquated glass tube, or Bosch-type, fuses. If a switch is faulty, that is not a serious problem.

Pull hard on the red lever and the vacuum-operated central locking system will open all the doors. It is intended for emergencies only.

Central locking system

4 3 2 1

Until the end of face-lift one production, the central locking was operated by a vacuum system. You can recognise this by the fact that all four doors never actually open and close at exactly the same time, there is always a momentary delay. The main thing is that they open and close! The boot (trunk) or tailgate and fuel filler flap are also locked centrally. If one door remains shut or opens only after some hesitation, it is usually the result of pressure loss in the system. Often, it is just a hose which is leaking or has come away. It is quite rare for the vacuum pump to be faulty.

Tailgate closing aid (estates)

4 3 2 1

Originally fêted as an up-market luxury feature, the electric closing aid, which automatically latches the tailgate shut, can be a right pain in its old age. Using it incorrectly (for example, slamming the tailgate shut) can cause the cable to break, triggering the little electric motor on the slightest bump in the road or when the car is parked. After just one night, the battery will be flat. During the test drive, be sure to listen out for a buzzing sound from the rear of the car.

Glass (stone chips, milky edges)

4 3 2 1

First and foremost, check for damage to the windscreen. Stone chips and cracks in the driver's field of view will not be accepted during MoT and other tests. Milky edges to the front or rear screens generally indicate leaks in the surrounding frame, which you should get to the bottom of in the short or medium term. Official testers will still often accept a narrow white band, but it will be noted as a minor fault.

Wheels and tyres

4 3 2 1

Check above all that any aftermarket wheels are officially authorised for fitment to the car. Ask to see the paperwork (certification). For the tyres, check their age (DOT number), tread depth and wear pattern. If they are unevenly worn, it may be necessary to adjust the alignment. Take a look at the outer edge of the wheel rims: is there any kerbing damage?

Wheels don't often look as good as this one. Small scratches aren't bad, but damage to the flange can affect safety.

With the 500E/E500, excessive wear on the inside edge of the rear tyres may be due to problems with the self-levelling suspension: if you suspect this, be sure to have the system thoroughly inspected.

Wheelarches and liners

4 3 2 1

The best way is to take a torch and shine it into all four wheelarches. Are all four liners present? How do the areas around the rubber plugs inside the wheelarches look? Traces of rust are a warning sign!

Wheel bearings, propshaft and ball joints

If there is any damage to any of these parts, you will find out at the latest during a test drive. Faulty wheel bearings, ball joints and propshafts all draw attention to themselves by the noise they make. Screeching or grinding noises, especially when cornering, give notice that one or more wheel bearings will soon have had it. Ball joints moan and groan terribly as the front suspension compresses. Repairs are a matter of routine and the parts are very affordable, but the job should not be put off.

Self-levelling suspension

One failure, which predominantly affects the estates (and more rarely the saloons, for which the self-levelling suspension was also available), is that of the pressure accumulators for the rear axle. If the back of the estate stands noticeably high and the rear suspension bounces during the test drive, then the pressure accumulators are nearly always faulty, and hardly ever the shock absorbers. Replacements are available from independent parts suppliers and from brands such as Lemförder (Mercedes OEM supplier) and Bilstein: they are inexpensive, as is the job of fitting them by a garage. It will be pricey if the shock absorbers are replaced at the same time, so check carefully if that is really necessary.

Steering

The 124 series cars are fitted with recirculating ball power steering. This is very light in use, but a rather indirect steering feel is normal. As the cars grow older, the free play in the steering will increase, but should not do so excessively. A professional garage can re-adjust the steering to a certain extent.

On the 500E/E500, the steering pump may suffer from wear after 100,000 miles (160,000km), often when a seal fails.

The huge wheel is pleasantly light to turn, though nearly always develops some free play over time. (Courtesy Frank Homann)

Brake callipers and pads 4 3 2 1

When examining the front brakes, turn the front wheels from side to side. You will then have a relatively unobstructed view of the brake callipers and pads. These should show a minimum thickness of 2-3mm, which in practice corresponds to the so-called wear limit for driving.

Brake discs (rotors) 4 3 2 1

You'll only get a good impression of the brake discs by removing the wheels. When you first examine your chosen car, however, you should at least check through the openings in the wheel whether the disc feels warped or scored. Run your fingertips along the edge of the disc. The more you can feel a clearly raised edge, the more urgent it is to replace the brake disc.

Foot-operated parking brake 4 3 2 1

The parking brake in the footwell often suffers from neglect, especially on cars with automatic transmission. Once the selector is in 'P,' hardly anyone uses it. Later, that can come back to haunt you, when it seizes up and the test inspector finds fault with its inadequate performance. So check whether it can be released with a clearly audible mechanical sound, that the pedal travel is not too long and, above all, that it will hold the car.

The hinges of the large bonnet are extremely strong, but you can easily catch your fingers on the locking lever.

Under the bonnet (hood) 4 3 2 1

Open the bonnet and make sure that it can be secured in both possible opening positions. Check whether all components are in original condition or if it has been spoilt with accessory parts from third-party suppliers (performance air filters etc).

Are there noticeable signs of oil or white stains, which might suggest that the coolant has boiled over or leaked out? Does the accessory drivebelt look good? Are there any conspicuous noises when the engine is running (clattering or rattling)? Don't be taken in by a freshly cleaned engine! An engine which is getting on in years should look like it's done some work, but be free from leaks.

On the later six-cylinder petrol-engined cars (280E/E280 and 320E/E320), look carefully at the engine wiring harness: the sheathing on this degrades over time, causing misfires. Mercedes offers a repair loom, which will fix the problem in the majority of cases, and costs far less than replacing the entire harness.

On the 500E/E500, the big V8 is a tight fit and may suffer from overheating. Signs of overheating on the firewall are not uncommon, and are not necessarily a concern.

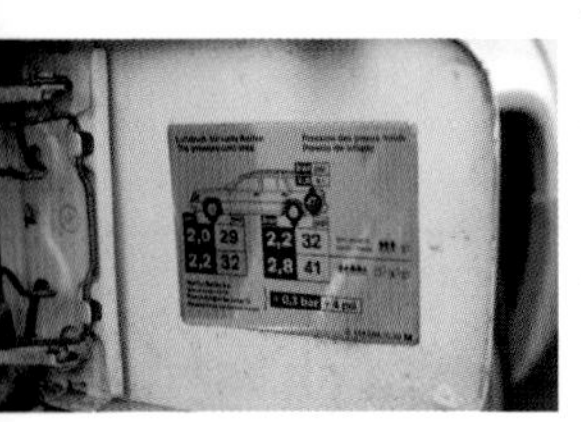

Important information at the ready inside the filler flap. The sticker should be in place.

Fuel filler flap and luggage compartment 4 3 2 1

It's worth taking a look behind the fuel filler flap (which hopefully unlocks cleanly), as the rot can also set in there due to a lack of care. The boot lid should rise up cleanly and smoothly and close effortlessly. On the estate, the gas struts

A spare wheel like this won't help you any more if you get a puncture. Get rid of the ancient tyre!

should be powerful enough to hold the heavy tailgate fully open. What is your impression of the luggage compartment? Is it clean? Is the (very high-quality) carpeting free from damage? Check that the tool kit, warning triangle and first aid kit are complete and in good condition; it's also worth looking at the spare wheel. You can get rid of an outdated or completely worn out tyre right away. In addition, keep a lookout for stains and signs of water getting in.

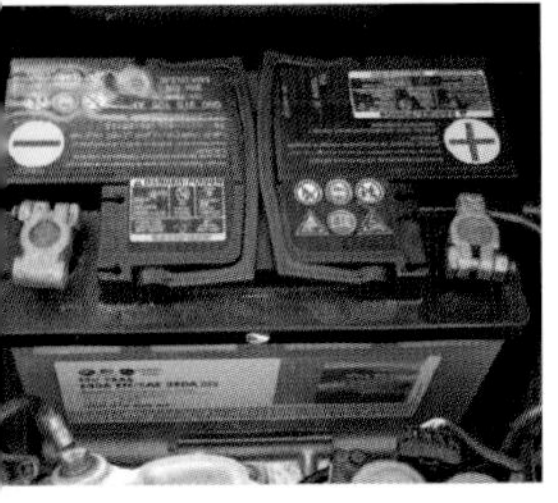

Check carefully that the battery fitted is really suitable for the car. Diesels need significantly more current than petrol engines to start.

Battery   

In order to check the condition of the battery, you will need a multimeter. When the meter is connected to the battery post (which is hopefully free from corrosion), it should give a reading of between 13.8 and 14.2 volts with the engine idling, and from 12.4 to 12.6 volts with the engine switched off. If the reading with the engine off is lower than 12.4 or even 12 volts, the battery should be changed.

Water getting in   

Check throughout the interior (carpets and seats) for water marks. Feel for signs of dampness inside all the door pockets and lift the carpets in the front footwells. Testing to see if a handkerchief absorbs any moisture will give you a further clue.

Interior

It looks hideous, but a skilled trimmer can repair it.

The fabrics and materials used inside the W124 are also of excellent quality. All the more reason to deduct points if you find tears, stains or worn switchgear. Check the seats, carpets, trim panels, buttons, switches, steering wheel, gearlever and headlining. Does it smell musty? Is all the equipment original, or has it been changed by the previous owner(s)? Check the steering wheel, seats and audio system. The side bolsters of the driver's seat should only be worn out in heavily-used W124s. Examine the adjustment mechanism for the seats (manual and electric), and check that all the switches and buttons function. Does the interior lighting work, in the front and back, and (on estates) in the C and D pillars? Can the rear head restraints be raised and folded away, and then flipped back using the switch on the dashboard?

Multi-function column stalk   

You should confirm that all the functions of the column stalk, which operates the headlamp main beam, windscreen washer and wipers as well as the indicators, work correctly. Replacements aren't cheap.

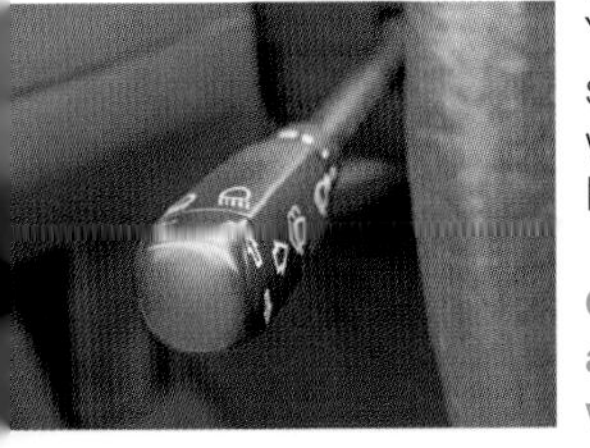

Check that all the functions work. There are hardly any used parts available in good working order.

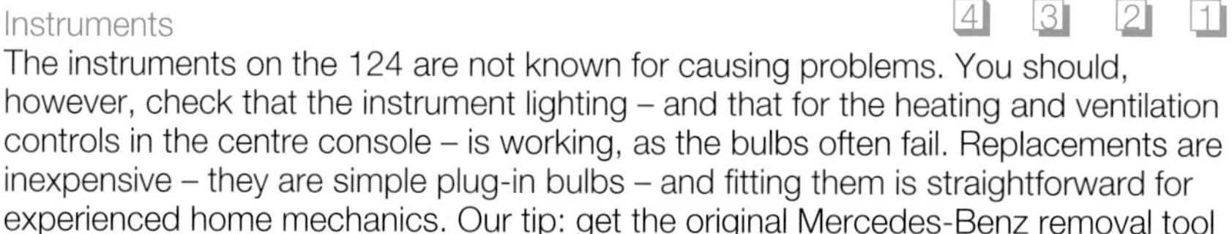

Instruments

4 3 2 1

The instruments on the 124 are not known for causing problems. You should, however, check that the instrument lighting – and that for the heating and ventilation controls in the centre console – is working, as the bulbs often fail. Replacements are inexpensive – they are simple plug-in bulbs – and fitting them is straightforward for experienced home mechanics. Our tip: get the original Mercedes-Benz removal tool for the instrument cluster, with a T-handle.

Warning lights

4 3 2 1

The number of warning lights depends on the specification of the car. You can read in the owner's manual what they signify. When you turn on the ignition, all the lights should come on. On diesel models, check the glow plug warning light on the far right, which should go out after a few seconds. One quirk of the 124 is that the ABS warning light coming on doesn't necessarily point to a fault in the system. Sometimes it's just that the voltage is too low. Hold on, and switch the engine off and on again. If the ABS light goes out after a few moments, the anti-block braking system is okay. Of course, though, it does no harm to have the fault examined.

Suspension

4 3 2 1

The suspension is considered to be robust. Caution is advised if a car has been modified, for example if the suspension has been lowered, or extremely wide tyres fitted. In this case, you should get the suspension checked over in a garage.

Exhaust system

4 3 2 1

You will be able to see and hear leaks in the exhaust system. Fortunately, replacements are affordable. If modifications (such as fitting a performance exhaust) have been made, you should make sure that the changes have been approved and correctly carried out. Ask to see the paperwork.

Oil leaks

4 3 2 1

It's easiest to check if a car is free from oil leaks after a test drive. When you are inspecting the engine, gearbox and differential, a mirror can be helpful. Place some cardboard under the car and look for oil dripping onto it.

Is the washer fluid topped up and to the right strength?

Is the coolant dark in colour, does it smell odd, or are there streaks of oil floating on top of it?

This hydraulic fluid reservoir was only fitted to cars with self-levelling suspension (estates and, as an option, saloons). The fluid level should be checked regularly. NB Use only the approved fluid type!

Coolant leaks 4 3 2 1

See if any coolant is seeping out under the front of the car. Light discolouration of the external components of the cooling system (hoses and clamps) may indicate leaks.

Fluids 4 3 2 1

Check the age and condition of all fluids, ideally referring to any service documents and stickers in the engine compartment. If any of them – with the exception of the engine oil – are noticeably dark, they should definitely be changed. And, of course, they should be filled to the correct level.

Test drive 4 3 2 1

Under no account let yourself be talked out of taking a decent test drive. There are dealers, but also private sellers, who will try to stop you with dubious excuses, or put you under time pressure: "Everything's OK with it," "You can drive it if you buy it afterwards." Drive the car in as many different conditions as you can, in town, on a country road and on the motorway (freeway). Test all the controls, run through the gears, apply the brakes firmly on a clear section of road. And, above all, look, listen, feel and smell! Listen very carefully, if something seems suspicious. Does anything smell unusual? Look at all the instruments and warning lights. Go by the seat of your pants: how does the car feel and sit on the road?

Making really sure 4 3 2 1

A former owner who is confident in his car will certainly have no objection if you want to have the car checked over by an independent inspector. Many garages, car clubs and vehicle testing organisations offer used car inspections. Naturally, these aren't free of charge, and the seller is unlikely to accept these costs. You should therefore feel pretty certain that the model you are looking at really is your next car. But the bottom line is that a professional assessment like this is a win for both parties.

Evaluation procedure

Add up the points scored. 136-152 points = outstanding car, which is sure to increase in value. 114-135 points = good to very good, you can buy this car with no reservations. 90-114 points = average to good, but what are the problems you have found? 76-90 points = below average to fair. Think carefully about the purchase. 60-76 points = usable as a winter car or until the next MoT or official inspection. Some expenditure will be required. If the price is particularly good or you have some mechanical skills, you could take the risk; otherwise, better not. 60 or fewer points = at best, a spares car. The complete restoration of a W124 is not (yet) worthwhile, there are enough good cars.

10 Auctions

– sold! Another way to buy your dream

Auction pros & cons

Pros: Prices are often lower than those of dealers or private sellers, and you might grab a real bargain on the day. Auctioneers have usually established clear title with the seller. At the venue you can usually examine documentation relating to the vehicle.
Cons: You have to rely on a sketchy catalogue description of condition & history. The opportunity to inspect is limited and you cannot drive the car. Auction cars are often a little below par and may require some work. It's easy to overbid. There will usually be a buyer's premium to pay in addition to the auction hammer price.

Which auction?

Auctions by established auctioneers are advertised in car magazines and on the auction houses' websites. A catalogue, or a simple printed list of the lots for auctions might only be available a day or two ahead, though often lots are listed and pictured on auctioneers' websites much earlier.

Contact the auctioneers to ask if previous auction selling prices are available as this is useful information (details of past sales are often available on websites).

Catalogue, entry fee, and payment details

When you purchase the catalogue of vehicles in the auction, it often acts as a ticket allowing two people to attend the viewing days and the auction. Catalogue details tend to be comparatively brief, but will include information such as 'one owner from new, low mileage, full service history,' etc. It will also usually show a guide price to give you some idea of what to expect to pay and will tell you what is charged as a 'Buyer's premium.' The catalogue will also contain details of acceptable forms of payment. At the fall of the hammer an immediate deposit is usually required, the balance payable within 24 hours. If the plan is to pay by cash, there may be a cash limit. Some auctions will accept payment by debit card. Sometimes credit or charge cards are acceptable, but will often incur an extra charge. A bank draft or bank transfer will have to be arranged in advance with your own bank as well as with the auction house. No car will be released before *all* payments are cleared. If delays occur in payment transfers, then storage costs can accrue.

Buyer's premium

A buyer's premium will be added to the hammer price: *don't* forget this in your calculations. It is not usual for there to be a further state tax or local tax on the purchase price and/or on the buyer's premium.

Viewing

In some instances, it's possible to view on the day, or days before, as well as in the hours prior to, the auction. There are auction officials available who are willing to help out by opening engine and luggage compartments and to allow you to inspect the interior. While the officials may start the engine for you, a test drive is out of the question. Crawling under and around the car as much as you want is permitted, but you can't suggest that the car you are interested in be jacked up, or attempt to do the job yourself. You can also ask to see any documentation available.

Bidding

Before you take part in the auction, *decide your maximum bid – and stick to it!*

It may take a while for the auctioneer to reach the lot you are interested in, so use that time to observe how other bidders behave. When it's the turn of your car, attract the auctioneer's attention and make an early bid. The auctioneer will then look to you for a reaction every time another bid is made, usually the bids will be in fixed increments until the bidding slows, when smaller increments will often be accepted before the hammer falls. If you want to withdraw from the bidding, make sure the auctioneer understands your intentions – a vigorous shake of the head when he or she looks to you for the next bid should do the trick!

Assuming that you are the successful bidder, the auctioneer will note your card or paddle number, and from that moment on you will be responsible for the vehicle.

If the car is unsold, either because it failed to reach the reserve or because there was little interest, it may be possible to negotiate with the owner, via the auctioneers, after the sale is over.

Successful bid

There are two more items to think about: how to get the car home, and insurance. If you can't drive the car, your own or a hired trailer is one way, another is to have the vehicle shipped using the facilities of a local company. The auction house will also have details of companies specialising in the transfer of cars.

Insurance for immediate cover can usually be purchased on site, but it may be more cost-effective to make arrangements with your own insurance company in advance, and then call to confirm the full details.

eBay & other online auctions

eBay & other online auctions could land you a car at a bargain price, though you'd be foolhardy to bid without examining the car first, something most vendors encourage. A useful feature of eBay is that the geographical location of the car is shown, so you can narrow your choices to those within a realistic radius of home. Be prepared to be outbid in the last few moments of the auction. Remember, your bid is binding and that it will be very, very difficult to get restitution in the case of a crooked vendor fleecing you – *caveat emptor!*

Be aware that some cars offered for sale in online auctions are 'ghost' cars. *Don't* part with *any* cash without being sure that the vehicle does actually exist and is as described (usually pre-bidding inspection is possible).

Auctioneers

Barrett-Jackson www.barrett-jackson.com
Bonhams www.bonhams.com
British Car Auctions (BCA) www.bca-europe.com/www.british-car-auctions.co.uk
Cheffins www.cheffins.co.uk
Christies www.christies.com
Coys www.coys.co.uk
eBay www.eBay.com
H&H www.classic-auctions.co.uk
RM www.rmauctions.com
Shannons www.shannons.com.au
Silver www.silverauctions.com

11 Paperwork

– correct documentation is essential!

The paper trail

Classic, collector and prestige cars usually come with a large portfolio of paperwork accumulated and passed on by a succession of proud owners. This documentation represents the real history of the car and from it can be deduced the level of care the car has received, how much it's been used, which specialists have worked on it and the dates of major repairs and restorations. All of this information will be priceless to you as the new owner, so be very wary of cars with little paperwork to support their claimed history.

Registration documents

All countries/states have some form of registration for private vehicles whether it's like the American 'pink slip' system or the British 'log book' system. It is essential to check that the registration document is genuine, that it relates to the car in question, and that all the vehicle's details are correctly recorded, including chassis/VIN and engine numbers (if these are shown). If you are buying from the previous owner, his or her name and address will be recorded in the document; this will not be the case if you are buying from a dealer.

In the UK the current (Euro-aligned) registration document is the 'V5C,' printed in three main coloured sections, blue, green and pink. The blue section relates to the car specification, the green one has details of the new owner, and the pink section is sent to the DVLA in the UK when the car is sold. A small section in yellow deals with selling the car within the motor trade. Upon payment of a small fee, the DVLA will provide details of earlier keepers of the vehicle – much can be learned in this way.

If the car has a foreign registration, there may be expensive and time-consuming formalities to complete. Do you really want the hassle?

Roadworthiness certificate

Most country/state administrations require that vehicles are regularly tested to prove that they are safe to use on the public highway and do not produce excessive emissions. In the UK that test (the 'MoT') is carried out at approved testing stations, for a fee. In the USA the requirement varies, but most states insist on an emissions test every two years as a minimum, while the police are charged with pulling over unsafe-looking vehicles.

In the UK the test is required on an annual basis once a vehicle becomes three years old. Of particular relevance for older cars is that the certificate issued includes the mileage reading recorded at the test date. It is therefore an independent record of that car's history. Ask the seller if previous certificates are available. Without an MoT the vehicle should be trailered to its new home, unless you insist that a valid MoT is part of the deal. (Not such a bad idea, as at least you will know the car was roadworthy on the day it was tested, and you don't need to wait for the old certificate to expire before having the test done.)

Road licence

Every country/state charges a tax for the use of its road system, the form of which, and how it is displayed, varies widely country to country and state to state.

Whatever the form of the 'road licence,' it must relate to the vehicle carrying it, and must be present and valid if the car is to be driven on the public highway legally. The value of the licence will depend on the length of time it will continue to be valid.

Changed legislation in the UK means that the seller of a car must surrender any existing road fund licence, and it is the responsibility of the new owner to re-tax the vehicle at the time of purchase and before the car can be driven on the road. It's therefore vital to see the Vehicle Registration Certificate (V5C) at the time of purchase, and to have access to the New Keeper Supplement (V5C/2), allowing the buyer to obtain road tax immediately.

If the car is untaxed because it has not been used for a period of time, the owner has to inform the licensing authorities, otherwise the vehicle's date-related registration number will be lost and there will be a painful amount of paperwork to get it re-registered.

Certificates of authenticity

For many makes of collectible car, it is possible to get a certificate proving the age and authenticity (eg engine and chassis numbers, paint colour and trim) of a particular vehicle. These are sometimes called 'Heritage Certificates' and if the car comes with one of these it is a definite bonus. If you want to obtain one, the relevant owners' club is the best starting point.

If the car has been used in European classic car rallies it may have a FIVA (Fédération Internationale des Véhicules Anciens) certificate. The so-called 'FIVA Passport,' or 'FIVA Vehicle Identity Card,' enables organisers and participants to recognise whether or not a particular vehicle is suitable for individual events. If you want to obtain such a certificate go to www.fbhvc.co.uk or www.fiva.org; there will be similar organisations in other countries too.

Valuation certificate

Hopefully the vendor will have a recent valuation certificate, or letter signed by a recognised expert stating how much he, or she, believes the particular car to be worth (such documents, together with photos, are usually needed to get 'agreed value' insurance). Generally, such documents should act only as confirmation of your own assessment of the car rather than a guarantee of value as the expert has probably not seen the car in the flesh. The easiest way to find out how to obtain a formal valuation is to contact the owners' club.

Service history

Often these cars will have been serviced at home by enthusiastic (and hopefully capable) owners for a good number of years. Nevertheless, try to obtain as much service history and other paperwork pertaining to the car as you can. Naturally, dealer stamps, or specialist garage receipts score most points in the value stakes. However, anything helps in the great authenticity game, items like the original bill of sale, handbook, parts invoices and repair bills, adding to the story and the character of the car. Even a brochure correct to the year of the car's manufacture is a useful document and something that you could well have to search hard to locate in future years. If the seller claims that the car has been restored, then expect receipts and other evidence from a specialist restorer.

If the seller claims to have carried out regular servicing, ask what work was

completed, when, and seek some evidence of it being carried out. Your assessment of the car's overall condition should tell you whether the seller's claims are genuine.

Restoration photographs

If the seller tells you that the car has been restored, then expect to be shown a series of photographs taken while the restoration was under way. Pictures taken at various stages, and from various angles, should help you gauge the thoroughness of the work. If you buy the car, ask if you can have all the photographs as they form an important part of the vehicle's history. It's surprising how many sellers are happy to part with their car and accept your cash, but want to hang on to their photographs! In the latter event, you may be able to persuade the vendor to get a set of copies made.

12 What's it worth?

– let your head rule your heart

Fantasy and reality

The 124 series stands right on the cusp of being just used cars or classics. Early models already qualify for classic status, while thousands of later cars (by which we refer to both the first and second face-lift models) go untroubled about their daily business as hard-used secondhand cars – a situation which many fortune-hunters – dealers and private sellers alike – try to take advantage of when setting their price. It's only to be expected, you can cash in on the 'Classic' label. A moderately well-equipped 230E saloon (sedan) in decent condition (ie requiring no immediate work), and with a clear history, will never justify a five-figure selling price! Nonetheless, offers like this frequently come up on the market. What matters when you're looking for a new set of wheels is to keep a cool head and not let your heart or gut feeling make the decision for you – even if the supposed car of your dreams is standing in front of you!

There's no such thing as a fully equipped car

Many sellers base their exaggerated asking price on the notion that the car in question is a rare 'fully loaded' model. In truth, no such car exists! Even at a time when Mercedes had yet to develop its equipment lines and packages, and new car buyers could indeed equip their cars on an individual basis, it was never possible to combine all the available options shown in the brochure in a single car. Some options could not be combined with all engines or fitted to all body styles. If a seller spins you the 'fully equipped story' (which unfortunately happens nearly all the time in the world of classic Mercedes), you can quietly smile and ignore it.

Be patient and you'll find the car of your dreams. Don't let yourself be blinded by over-enthusiasm! (Courtesy Frank Homann)

Condition

Just one thing matters: what is standing in front of you? With the help of the rating system in Chapter 9, you can now assess the condition of the car on offer and whether it is worthwhile to start negotiating with the seller. In addition, it's sensible to find out about the current market situation beforehand (eg through the relevant websites) and to look at specialist magazines such as *Mercedes Enthusiast* or *Classic Mercedes* (see Chapter 16, The Community). In the case of Mercedes-Benz, it's always worthwhile getting in touch with the lively enthusiast community. Clubs will often be very helpful, even to non-members. As a rule, you can get a pretty good idea of the going price from sources like these. Obviously, you can pay correspondingly more for a really well cared-for car with one pensioner owner, that has been kept in a garage. And such cars really do exist! But you should be aware that the asking price will depend not only on the condition *per se*, but on the extras that are fitted.

The electric rear sunblind is a useful extra ... if it's still working. (Courtesy Frank Homann)

Desirable options/extras

Automatic transmission
Sliding roof
Velour or leather upholstery
Orthopaedic driver's seat
Original radio (Becker)
Metallic paint
Original alloy wheels (such as the classic 15-hole design)
Air-conditioning
Tinted glass
Catalytic converter (petrol engines)

Some seats have a memory function: this feature was often ordered. Check that it works correctly. (Courtesy Frank Homann)

Undesirable features

Unapproved or non-standard accessories (performance tuning)
Non-standard paint colours
'Wrong' colours
Basic specification, without any extras

Exceptions

The so-called 'wrong' colours are those finishes which were admittedly available from the factory, but that were rarely ordered. The iridescent turquoise 'Beryl' metallic, the unusual 'Impala' brown or the not uncommon 'Almandine' red are generally colours that justify a price reduction, as they are reputedly hard to sell. Recently though, collectors are emerging in the 124 community who are specifically looking for such cars. Even 'stripped out' basic cars, the so-called 'book-keeper spec' models, which are inherently unpopular, are becoming more and more attractive among enthusiasts – with the motto: "If it isn't there, it can't go wrong."

Warranties

The matter of warranties on older cars is always difficult. Even when you buy from a dealer, who is obliged to offer a warranty, there will be times when some traders try to get out of it. There can be sales agreements with dubious disclaimers, which release the trader from any liability and transfer all the risk to the buyer. In such cases it may often be said that the car is 'intended for export.' If talk like that makes you feel uneasy, take your leave of the seller and look around for a more serious dealer. Mind you, when you buy from a private seller, you always bear the risk yourself!

Pre-purchase inspection

To keep the risk to a minimum, first, you have this book in your hand, and secondly, you can call on the help of independent experts. A seller who has nothing to hide should agree to a technical inspection at a garage of the car you are considering buying (it doesn't have to be a Mercedes dealership), or by a vehicle testing organisation. You will have to bear the costs of this, which is normally worthwhile. If the seller is sceptical, you can argue that he will be better able to evaluate his car after a professional inspection. It's really a win-win for everyone involved.

Striking a deal

Most importantly, insist on a written sales agreement! Striking a deal on a handshake has its charm, but definitely not between two strangers. Negotiate the price on the basis of the actual condition of the car, its mileage, the service documentation available and any faults. Take into account the specification and body style: at present convertibles and estates are worth more than saloons and coupés. If you can see that immediate repairs or service work will be required, deduct these from the purchase price or negotiate with the seller to what extent he can repair these faults before the sale. Always be fair! Part of every negotiation is to meet each other in the middle and to reach a price which both parties can live with. If that just isn't possible, keep looking! There are so many well-maintained 124s on the market that you will find a decent car in any case.

13 Do you really want to restore?

– it'll take longer and cost more than you think

There's a reason why many 124s that are no longer fit for service are exported to Africa: apart from the fact that it is a sturdily built car, the mechanical parts can be easily understood and serviced. Of course, depending on its specification, a 124 has some electronic components, which demand know-how and suitable tools to service and repair, but on the mechanical side, any experienced repairman should be able to keep a W124 running. As a master mechanic from Morocco put it, "With the gift for improvisation which the youngsters in my country have, an old 'Merc' like this will run forever."

Despite its very sturdy construction, dismantling, repairing and reassembling the bodywork is by no means insurmountable for those with experience, especially as there are enough replacement parts – new and used – available on the market. There is no need to resort to reproduction parts, as most items can still be found in the parts stores of Mercedes dealerships. It's also possible to save money with the extensive stock at Mercedes' own Used Parts Centre (MBGTC), while a search in the right breakers' yards or on the internet can likewise be promising. You should be wary of cars which show signs of previous repairs, or which were simply fixed up to get through an official inspection. You'll always pay dearly for sloppy work.

As far as electrical and electronic components are concerned, there is a good selection of replacement parts in all price ranges on the secondhand market, above all at the MBGTC (with guaranteed parts that have been checked). Beware, though, of offers which are far too cheap: you can't check that an ABS control unit is working on the kitchen table, so only buy parts like this from trustworthy suppliers.

You can buy a car like this and drive it every day. Restoring it isn't worthwhile.

Questions and answers

Is it worth restoring a car such as the Mercedes W124 which is still found in such numbers on the road today? The author is inclined to answer, "No." Why? First, in practice there are still enough good cars on the market – and in all the body styles and many different specifications – that in case of doubt it's better to spend your time looking for a better car than sink your money into a run-down example. Secondly, prices for the 124 – with the exception of very well-maintained convertibles, the exclusive 500E/E500 or the rare AMG versions – have not yet reached the exalted levels at which a complete restoration would ultimately help maintain or increase the car's value.

In most cases, the thorough overhaul of a 124 – especially if carried out by a professional garage – will not be worth it. In any event, the costs would exceed the value of the car. That can and will change, but for the moment that's the situation. Unless the vehicle in question is a very rare model or there is a personal story involved ("It was Grandpa's 'Merc,' I absolutely must keep it!"), you should steer clear of a complete restoration on financial grounds.

In this case, a thorough overhaul is at least worth considering: a 300TE 4Matic is rare. It is in reasonable, but not perfect, condition. (Courtesy Frank Homann)

14 Paint problems

– bad complexion, including dimples, pimples and bubbles

Paint faults generally occur due to lack of protection/maintenance, or to poor preparation prior to a respray or touch-up. Some of the following conditions may be present in the car you're looking at.

Orange peel

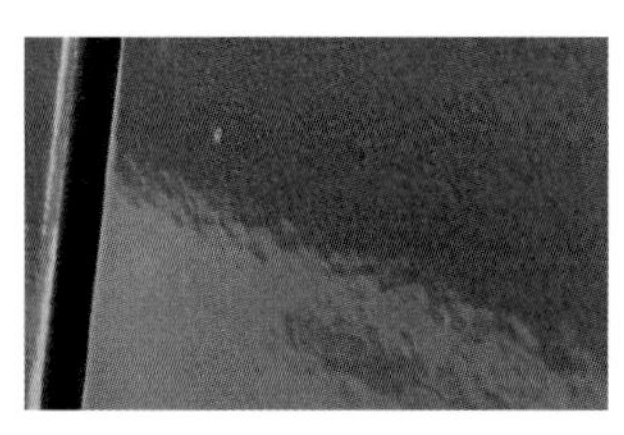

This appears as an uneven paint surface, similar to the appearance of the skin of an orange. The fault is caused by the failure of atomized paint droplets to flow into each other when they hit the surface. It's sometimes possible to rub out the effect with proprietary paint cutting/rubbing compound or very fine grades of abrasive paper. A respray may be necessary in severe cases. Consult a bodywork repairer/paint shop for advice on the particular car.

Cracking

Severe cases are likely to have been caused by too heavy an application of paint (or filler beneath the paint). Also, insufficient stirring of the paint before application can lead to the components being improperly mixed, and cracking can result. Incompatibility with the paint already on the panel can have a similar effect. To rectify the problem, it is necessary to rub down to a smooth, sound finish before respraying the problem area.

Crazing

Sometimes the paint takes on a crazed rather than a cracked appearance when the problems mentioned under 'Cracking' are present. This problem can also be caused by a reaction between the underlying surface and the paint. Paint removal and respraying the problem area is usually the only solution.

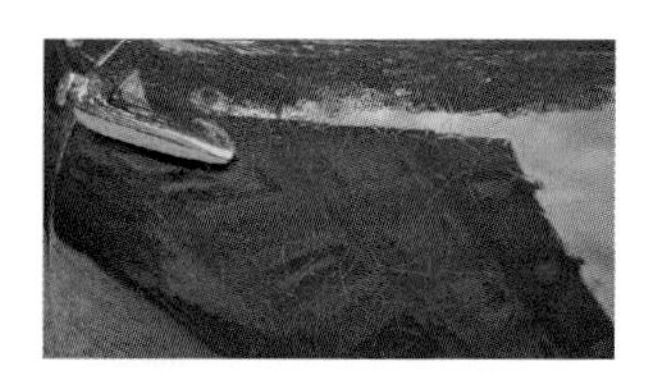

Blistering

Almost always caused by corrosion of the metal beneath the paint. Usually perforation will be found in the metal and the damage will usually be worse than that suggested by the area of blistering. The metal will have to be repaired before repainting.

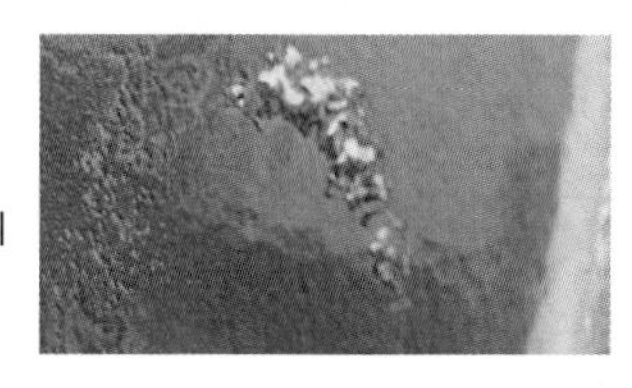

Micro blistering

Usually the result of an economy respray, where inadequate heating has allowed moisture to settle on the car before spraying. Consult a paint specialist, but usually damaged paint will have to be removed before partial or full respraying. Can also be caused by car covers that don't 'breathe.'

Fading

Some colours, especially reds, are prone to fading if subjected to strong sunlight for long periods without the benefit of polish protection. Sometimes proprietary paint restorers and/or paint cutting/rubbing compounds will retrieve the situation. Often a respray is the only real solution.

Peeling

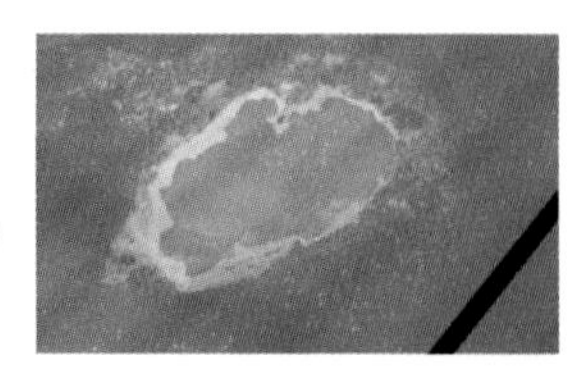

Often a problem with metallic paintwork when the sealing lacquer becomes damaged and begins to peel off. Poorly applied paint may also peel. The remedy is to strip and start again!

Dimples

Dimples in the paintwork are caused by the residue of polish (particularly silicone types) not being removed properly before respraying. Paint removal and repainting is the only solution.

Dents

Small dents are usually easily cured by the 'Dentmaster,' or equivalent process, that sucks or pushes out the dent (as long as the paint surface is still intact). Companies offering dent removal services usually come to your home: consult your telephone directory.

15 Problems due to lack of use

– just like their owners, W124s need exercise!

The best Mercedes is one which is regularly driven and maintained. However tempting the car with one retired owner which was kept in a garage and polished with a handkerchief, and covered only a certified 60,000km (37,000 miles), it may not be the best on offer! Later, when you want to use such a car on a regular basis, its condition may suffer from this period of immobilisation, resulting in costly repairs. On the club scene, cars with sometimes much higher mileages and a continuous service history are clearly favoured. Cars are for driving! If you use a W124 for high days and holidays only, you can, of course, consider a low mileage car.

Corrosion

Cars which have seen little of the road in their lives rarely suffer from body corrosion. For the most part, they spend their days and nights in dry garages. Instead, the 'brown plague' nibbles away unseen, preferring mechanical components like the braking system, where the pistons, callipers and cylinders can be affected, or where the brake pads can become seized onto the discs (rotors). Rust can also cause the parking brake mechanism to seize. It is not uncommon for the flywheel to stick to the clutch disc. It's better not to move a car like this until the problem has been fixed.

Fluids

Fluids which are too old should be replenished immediately before you begin to use a car which has been standing for a long time. Old oil can attack seals and gaskets, and damage bearings. Some constituents of fuel and brake fluid can absorb water from the air over time, which can lead to corrosion in the fuel lines and braking system. If the brake fluid has already absorbed too much water, vapour bubbles can form in the system when the brakes are hot. In the worst case, the brakes may fail!

The elaborately constructed rear axle ensures the imperturbable directional stability and safe handling of the W124. If the car is noticeably vague to drive, you should get the rear suspension inspected.

Tyres

It is a common misconception that tyres whose profile still looks quite alright can be used for ever without any worries. After just five to seven years, the softening agents in the tyre compound can lose their effectiveness, and the tyre will become brittle and cracked. That can put your life at risk! When you are buying the car, look at the DOT number moulded into the tyre wall. This gives the year and week of production. If the car has been standing for a long time, flat spots may have developed. They can be felt clearly as vibration when driving, although this can also disappear again. If it doesn't, you should have the tyres checked by a professional and, if necessary, replaced.

Shock absorbers

As a rule, shock absorbers rarely suffer from periods

of immobilisation. If they are suspiciously damp, however, oil may already be seeping out. In this case, the seals inside the shock absorbers may be leaking. Shock absorbers like this should be replaced. Carry out the well-known bounce test: if the car bounces up and down several times after you press down hard on one corner, the shock absorber has probably had it. Admittedly, this test won't tell you much more. Take care if you are looking at a car with sports suspension. Sometimes this can be so firm that there is no bounce anyway. In general, a car like this should be examined by an expert.

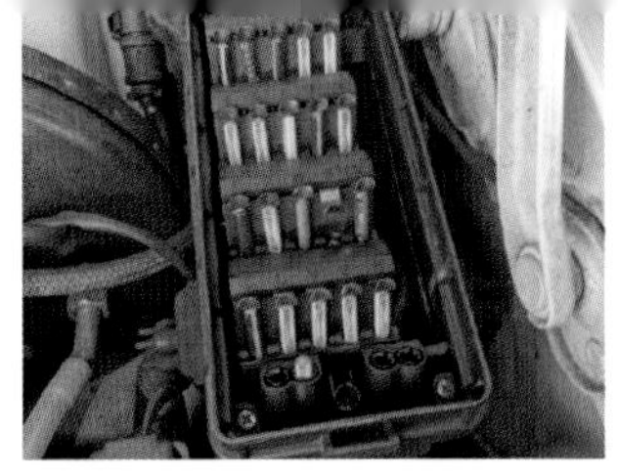

The antiquated Bosch-type fuses (up to and including the first face-lift models) are liable to work loose in their holders, causing failures of the in-car electrical systems.

Rubber and plastic parts

After more than 20 years, the softening agents in all kinds of rubber seals can lose their effectiveness and the seals will have hardened. Regular care with the appropriate sprays or roll-on pens will delay this process but cannot prevent it. A thorough check of all the seals concerned (doors, windows, sliding roof, etc) should go without saying. Faded bumpers, protection strips and mirror casings are more cosmetic in nature. Over time unpainted plastic parts become grey and unsightly. You can effectively counter these problems with the right care products. If you're unsure about doing it yourself, turn to a professional car detailer. It's often hard to believe what experts can do with clearly worn-out cars, and even paintwork. And it needn't be that expensive.

Electrics

The biggest enemy of correctly functioning electrical systems is oxidation. It can affect plug connectors and contacts, often leading to malfunctions or the outright failure of complete modules. If a car has been standing for a long time, you should thoroughly test all the important functions. If you do find a defect, there's no need to bring out your complete toolkit straightaway. In many cases, spraying the contacts or using some fine sandpaper on the connectors can work wonders. Even fuses can pack up, but can be replaced in no time. Normally, it will do your car good to connect the battery to a trickle charger (also known as a 'Battery Tender') if you expect it to be immobilised for a longer time. This simulates the charging and discharging process in daily use and so keeps the battery fresh.

This rear silencer is brand new and will last for many years. In fact, it's an aftermarket part (from Eberspächer) and was therefore available at a reasonable price from an independent parts supplier.

Exhaust system

It's almost a law of nature, and not just for the W124: when you purchase a car that hasn't moved for months, after a few weeks' use the exhaust system will give up. Even if it still looks intact, the water contained in the emissions will let rust develop in the silencer and tailpipe. They will rust from the inside out, it's almost inevitable – unless the car has been fitted with a stainless steel exhaust system, but even that is no guarantee against corrosion.

16 The Community

– key people, organisations and companies in the W124 world

Owners of classic Mercedes are well taken care of. Outstanding support from the factory, active clubs and a wide range of independent parts suppliers and specialists make it easy to maintain and look after a W124.

Clubs

Mercedes-Benz lends its official support to more than 80 clubs worldwide, which offer an extensive range of benefits and activities. These include discounts on parts and services, invitations to events and, of course, visits to the company's magnificent museum in Stuttgart. Find out more at:

- Mercedes-Benz Classic (factory homepage): www.mercedes-benz.com/en/mercedes-benz/classic/classic-overview/
- UK: www.mercedes-benz-club.co.uk/Home.aspx
- North America: www.mbca.org
- Other countries: specials.mercedes-benz-classic.com/en/club/#ger

Specialists

Many dealers in recent classics now specialise in selling the W124 series. In the UK, these include:

- Avantgarde Cars: www.avantgardecars.co.uk
- Charles Ironside: www.charlesironside.co.uk
- Cheshire Classic Benz (also servicing): www.ccbenz.co.uk
- Prestige Motor Company: www.prestigemotorcompany.co.uk
- Silver Arrows: www.silverarrows.co.uk
- W124 (specialists in estates, also offer an inspection service): www.w124.co.uk

There are many independent service specialists too, among them:

- Roger Edwards Motors: see Facebook page or call 01494 766 766
- John Haynes Mercedes: www.johnhaynesmercedes.co.uk
- Joseph Joos: www.josephjoos.co.uk
- Derrick Wells: www.derrick-wells.com
- D-class (specialist in interior & hood repairs): www.dclass.co.uk

You will find many more specialists who advertise in the magazines listed below.

Parts suppliers

Mercedes-Benz has an almost unrivalled reputation for the availability of parts for its older cars, including the W124, but prices can sometimes be high. Among UK dealers, Mercedes-Benz of Poole is well known, whilst Mercedes has its own Used Parts Center, the MBGTC (see below):

- Mercedes-Benz of Poole (formerly Jacksons): www.mercedes-benzofpoole.co.uk
- Mercedes Parts Centre: mercedes-parts-centre.co.uk
- PFS Parts: www.partsformercedes-benz.com
- Three Pointed Parts: www.threepointedparts.co.uk
- USA – Pelican Parts: www.pelicanparts.com/catalog/SuperCat/W124_catalog.htm
- Germany – MBGTC: www.mbgtc.de (currently in German language only).

Useful sources of information

Three English-language magazines cater to classic Mercedes enthusiasts and often feature the W124:

- *Mercedes Enthusiast* (monthly) and *Classic Mercedes* (quarterly) can be found at large newsstands in both the UK and North America, or obtained on subscription from www.mercedesenthusiast.co.uk
- *Mercedes-Benz Classic* is published in English and German by Mercedes itself three times a year: subscribe at www.mercedes-benz.com/en/mercedes-benz/lifestyle/mercedes-benz-magazines/classic-magazine/subscription/

As well as Mercedes' own manuals and technical literature, there are plenty of books for enthusiasts to read up on the 124 series cars. Brooklands Books (www.brooklandsbooks.co.uk) devotes one of its excellent compilations of period road tests to the model: *Mercedes-Benz E-Class W124 1985-1995*. The E36 and E60 AMG models are covered, too, in its road test collection, *Mercedes AMG 1983-1999*. Brooklands also publishes an *Owners' Workshop Manual* for all except the V8 models in the W124 series. The 400E and 500E models are, however, included in Bentley Publishers' *Mercedes-Benz E-Class (W124) Owner's Bible*. Finally, there is a good overview of the history of the 124 series in James Taylor's book, *Mercedes-Benz W124: The Complete Story* (Crowood Press).

Several internet sites help owners to share information about the cars and fix common problems:

- Mercedes-Benz Owners: www.mercedesclub.org.uk
- Mercedes Source (North America): mercedessource.com/chassis/124/
- W124-Zone: www.w124-zone.com

Back in the '80s, customers' cars would be received by the Mercedes dealer foreman, meticulously turned out with collar and tie under his overalls. Today, many specialists look after the 124 series. (Courtesy Daimler AG)

17 Vital statistics

– essential data at your fingertips

Technical data

W124 Four-cylinder petrol models (all bodystyles).
The 200 carburettor engine was not offered in the Coupé and Cabriolet.

Model	Model	200	200E
	Code	124.020	124.021
	Period of production	01/1985-06/1990	07/1985-10/1992 (until 09/1988 only for export to Italy)
Engine	Design: Four-stroke internal combustion	Four-cylinder inline M 102 V 20	Four-cylinder inline M 102 E 20
	Capacity (cc)	1997	
	Power output (bhp/rpm)	108/5200 (ECE); from 09/1986: 104/5500 (KAT)	120/5100 (ECE); from 09/1988 in Germany 116/5200 (KAT)
Wheels	Size	6 J x 15 H 2	6½ J x 15 H 2
	Tyres	185/65 R 15 87 H	195/65 R 15 91 H
Performance	Top speed in mph (km/h)	116 (187); KAT: 115 (185)	121 (195); KAT: 120 (193)
	Fuel consumption (mpg Imperial/mpg US)	31.7/26.4	31.4/26.1

W124 Six-cylinder petrol models (all bodystyles).
The Cabrio, Coupé and Estate were not available as the 260E, the Cabrio and Coupé not available as the 280E/E280. The E36 AMG was not available as a saloon.

Model	Model	260E	300E
	Code	124.026	124.030
	Period of production	09/1985-10/1992	04/1985-11/1992
Engine	Design: Four-stroke internal combustion	Six-cylinder inline M 103 E 26	Six-cylinder inline. M 103 E 30
	Capacity (cc)	2599	2962
	Power output (bhp/rpm)	168/5800 (ECE); from 09/1985: 164/5800 (RÜF) or 158/5800 (KAT)	187/5600 (ECE); from 09/1985: 185/5700 (RÜF) or 178/5700 (KAT); from 09/1989 KAT only
Wheels	Size	6½ J x 15 H 2	
	Tyres	195/65 VR 15; from 09/1989: 195/65 R 15 91 V	195/65 VR 15; from 09/1989: 195/65 R 15 91 V
Performance	Top speed in mph (km/h)	135 (218); KAT: 134 (215)	ECE: 143 (230); RÜF: 142 (228); KAT: 140 (225)
	Fuel consumption (mpg Imperial/mpg US)	28.8/24.0	27.4/22.8

W124 Eight-cylinder petrol models (only offered as saloons).

Model	Model	400E, from 07/1993: E420
	Code	124.034
	Period of production	05/1991-06/1995
Engine	Design: Four-stroke internal combustion	V8 M 119 E42
	Capacity (cc)	4196
	Power output (bhp/rpm)	275/5700
Wheels	Size	6½ J x 15; from 07/1993: 7J x 16
	Tyres	195/65 ZR 15; from 07/1993 215/55 ZR 16
Performance	Top speed in mph (km/h)	155 (250)
	Fuel consumption (mpg Imperial/mpg US)	23.9/19.9

	230E	200E, from 07/1993: E200	220E, from 07/1993: E220
	124.023	124.019	124.022
	01/1985-10/1992	09/1992-08/1995	09/1992-06/1996 (from 09/1995 only in CKD form for India)
	Four-cylinder inline M 102 E 23	Four-cylinder inline M 111 E 20	Four-cylinder inline M 111 E 22
	2298	1998	2199
	134/5100 (ECE); from 09/1985 also 130/5100 (KAT optional, from 09/1986 standard)	134/5500	148/5500
			195/65 R 15 91 V
	126 (203); KAT: 123 (200)	123 (200)	130 (210)
	31.4/26.1	32.8/27.4	32.1/26.7

	300E-24	280E, from 07/1993: E280	320E, from 07/1993: E320	E36 AMG
	124.031	124.028	124.032	124.092
	08/1989-10/1992	09/1992-08/1995	09/1992-08/1995	1993-1996
	Six-cylinder inline M 104 E 30	Six-cylinder inline M 104 E 28	Six-cylinder inline M 104 E 32	Six-cylinder inline M 104 E 36
	2960	2799	3199	3606
	217/6400	194/5500; from 07/1993: 190/5500	217/5500	268/5750
				7½J x 17
	195/65 ZR 15	195/65 R 15 91 V	195/65 ZR 15	225/45 ZR 17
	147 (237)	143 (230)	146 (235)	149 (240)
	25.9/21.6	26.4/22.0	25.7/21.4	25.7/21.4

	500E, from 07/1993: E500	E60 AMG
	124.036	124.036
	09/1990-04/1995	09/1993-11/1994
	V8 M 119 E50	V8 M 119 E60
	4973	5956
	322/5700 (KAT); from 09/1992 316/5600	376/5600
	8 J x 16	8 J x 16
	225/65 ZR 16	225/65 ZR 16
	155 (250)	155 (250)
	20.9/17.4	22.4/18.7

W124 Four and five-cylinder diesel models (all bodystyles).
The Coupé and Cabriolet were not offered with diesel engines.

Model	Model	200D, from 07/1993: E200 Diesel	250D
	Code	124.120	124.125
	Period of production	01/1985-08/1995	05/1985-07/1993
Engine	Design: Four-stroke diesel	Four-cylinder inline OM 601 D 20	Five-cylinder inline OM 602 D 25
	Capacity (cc)	1997	2497
	Power output (bhp/rpm)	71/4600; from 02/1989: 74/4600	89/4600; from 02/1989: 93/4600; from 09/1989 with optional catalytic converter: 89/4600
Wheels	Size	J x 15 H 2	6½ J x 15 H 2
	Tyres	185/65 R 15 87 T	195/65 R 15 91 T
Performance	Top speed in mph (km/h)	99 (160)	109 (175)
	Fuel consumption (mpg Imperial/mpg US)	38.2/31.8	39.8/33.1

W124 Six-cylinder diesel models (all bodystyles).
The Coupé and Cabriolet were not offered with diesel engines.

Model	Model	300D
	Code	124.130
	Period of production	01/1985-07/1993
Engine	Design (four-stroke diesel)	Six-cylinder inline OM 603 D 30
	Capacity (cc)	2996
	Power output (bhp/rpm)	108/4600; from 09/1989 with optional catalytic converter 109/4600
Wheels	Size	6½ J x 15 H 2
	Tyres	195/65 R 15 91 T; from 02/1989: 195/65 R 15 91 H
Performance	Top speed in mph (km/h)	118 (190)
	Fuel consumption (mpg Imperial/mpg US)	38.2/31.8

Chassis and running gear

- Chassis, body: all-steel body, unitary construction
- Front suspension: independent, with struts and wishbones
- Rear suspension: multi-link suspension, with optional self-levelling (standard on estates)
- Steering: recirculating ball, with power assistance
- Brakes: dual circuit with vacuum servo, discs front and rear, ABS optional (standard from 09/1988)

Transmission

- Gearbox: four- or five-speed manual with synchromesh on all gears, four- or five-speed automatic
- Drivetrain: rear-wheel drive, four-wheel drive (4Matic) optional

Dimensions

- Wheelbase – saloon and estate: 2800mm (110.2 inches)
- Wheelbase – coupé and cabriolet: 2715mm (106.9 inches)
- Length/width/height – saloon: 4740/1740/1446mm (186.6/68.5/56.9 inches)
- Length/width/height – estate: 4765/1740/1490mm (187.6/68.5/58.7 inches)
- Length/width/height – coupé: 4655/1740/1410mm (183.3/68.5/55.5 inches)
- Length/width/height – cabriolet: 4655/1740/1391mm (183.3/68.5/54.8 inches)

	250D Turbo, from 07/1993: E250 Turbodiesel	E250 Diesel
	124.128	124.126
	08/1988-06/1995	07/1993-06/1996 (from 09/1995 only in CKD form for Asian markets)
	Five-cylinder inline OM 602 D 25 A	Five-cylinder inline OM 605 D 25
	124/4600	111/5000
	195/65 R 15 91 H	195/65 R 15 91 V
	123 (198)	118 (190)
	37.2/30.9	39.8/33.1

	300D Turbo, from 07/1993: E300 Turbodiesel	E300 Diesel
	124.133	124.131
	04/1986-06/1995	07/1993-08/1995
	Six-cylinder inline OM 603 D 30 A	Six-cylinder inline OM 606 D 30
	141/4600; from 09/1988: 145/4600	134/5000
	195/65 R 15 91 H	
	126 (202)	124 (200)
	35.8/29.8	38.2/31.8

ECE = car not fitted with catalytic converter
RüF = car not fitted with catalytic converter, but can be retrofitted
Kat = car fitted with catalytic converter

The Essential Buyer's Guide™ series ...

... don't buy a vehicle until you've read one of these!

£9.99-£12.99 / $19.95-$25.00
(prices subject to change, p&p extra).

For more details visit www.veloce.co.uk or email info@veloce.co.uk

Also from Veloce Publishing ...

Mercedes-Benz SLK

R170 series 1996-2004
Brian Long

This book reveals the full history of the first generation Mercedes-Benz SLK, covering in detail the German, US, UK, Australian and Japanese markets. The perfect book to grace a Mercedes-Benz enthusiasts' library shelf, it's the definitive record of the model illustrated with stunning photographs.

ISBN: 978-1-845846-51-0
Hardback • 24.8x24.8cm
• £45* UK/$75* USA •
192 pages • 337 colour pictures

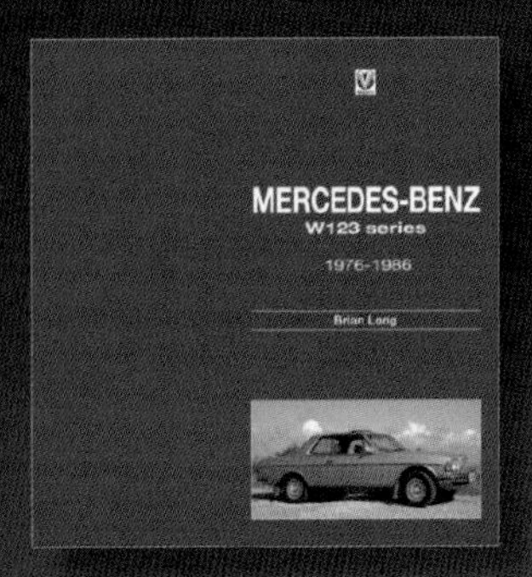

Mercedes-Benz W123 series

All models 1976 to 1986
Brian Long

The definitive history of the entire Mercedes-Benz W123 series. From the saloons/sedans, coupés, estates/wagons, to LWB and chassis only vehicles, this book contains an overview of all the models sold in each of the world's major markets. Packed full of information and contemporary illustrations sourced from the factory.

ISBN: 978-1-845847-92-0
Hardback • 25x25cm
• £45* UK/$75* USA •
192 pages • 321 colour pictures

*prices subject to change, p&p extra.
For more details visit www.veloce.co.uk or email info@veloce.co.uk